Breakdown of
Einstein's
Equivalence Principle

Breakdown of
Einstein's
Equivalence Principle

editor **Andrei G Lebed**
University of Arizona, USA

World Scientific

NEW JERSEY · LONDON · SINGAPORE · BEIJING · SHANGHAI · HONG KONG · TAIPEI · CHENNAI · TOKYO

Published by

World Scientific Publishing Co. Pte. Ltd.
5 Toh Tuck Link, Singapore 596224
USA office: 27 Warren Street, Suite 401-402, Hackensack, NJ 07601
UK office: 57 Shelton Street, Covent Garden, London WC2H 9HE

British Library Cataloguing-in-Publication Data
A catalogue record for this book is available from the British Library.

ISBN 978-981-125-358-4 (hardcover)
ISBN 978-981-125-359-1 (ebook for institutions)
ISBN 978-981-125-360-7 (ebook for individuals)

For any available supplementary material, please visit
https://www.worldscientific.com/worldscibooks/10.1142/12759#t=suppl

Typeset by Stallion Press
Email: enquiries@stallionpress.com

The editor dedicates his own contributions to the book to his wife, Natalia, for her loving support and patience.

Preface

From a historical point of view, the Equivalence Principle was first clearly formulated by Galileo Galilei (see Fig. 1) as the independence of a body's acceleration in the Earth's gravitational field on the body's mass. According to the popular legend, he dropped cannonballs of different sizes from the Leaning Tower of Pisa and found that they hit the ground at the same time. This was approximately 400–450 years ago. Using modern language, Galilei established the equivalence of inertial, m_i, and gravitational, m_g, masses with the accuracy about $|m_i - m_g|/m_i \leq 2 \times 10^{-2}$. Much less is known about his predecessor, Greek philosopher John Philoponus, who performed less sophisticated but similar experiments approximately 1500 years ago. Newtonian physics accepted Galileo's Equivalence Principle and the first successful theory of gravity, created by Isaac Newton, explained the main features of our planetary system, including the empirical Kepler laws.

In 1916, Albert Einstein (see Fig. 2) published his theory of General Relativity, which is based on Einstein's Equivalence Principle. Before he elaborated General Relativity, in 1911, he had formulated the Equivalence Principle in the following way. Einstein considered two reference frames, K and K', where K was placed in a uniform gravitational field. K' was not in a gravitational field, but it was uniformly accelerated. He wrote: "As long as we restrict ourselves to pure mechanical processes in the realm where Newton's mechanics holds sway, we are certain of the equivalence of the systems K and K'. But this view of ours will not have any deeper significance unless the systems K and K' are equivalent with respect to all physical processes, that is, unless the laws of nature with respect to K are in entire agreement with those with respect to K'. By assuming this to be so, we arrive at a principle which, if it is really true, has great heuristic importance. By theoretical consideration of processes which take

Fig. 1.   Galileo Galilei (1564–1642) (drawing of Natalia Lebed).

Fig. 2.   Albert Einstein (1879–1955) (drawing of Natalia Lebed).

place relatively to a system of reference with uniform acceleration, we obtain information as to the career of processes in a homogeneous gravitational field."

We note that modern physics distinguishes between three variants of the Equivalence Principle: weak Equivalence Principle, Einstein's Equivalence Principle, and strong Equivalence Principle. The weak Equivalence Principle is equivalent to the so-called universality of a free fall and is related to the original Galileo experiments. This variant has been confirmed with great accuracy for ordinary solid-state bodies. In the recent space mission, MICROSCOPE of the French space agency CNES, the experimental equipment consisted of two cylindrical shells: larger cylindrical shell and smaller one. The smaller shell, made of platinum and rhodium, was placed inside the larger one, made of titanium and aluminum. In more than 2 years and 5 months, physicists on the Earth observed a relative motion of the cylinders and found that they were accelerated with the same quantity with accuracy better than $|m_i - m_g|/m_i \leq 10^{-16}$. We can say that the Equivalence Principle is one of the best established laws in physics. Moreover, two more space missions are currently in their design phases: Satellite Test of the Equivalence Principle (STEP) and Galileo Galilei (GG). The STEP mission is being developed by an international team of collaborators under the leadership of Stanford University, whereas the GG is an Italian project. The missions would be operated in a similar way as the MICROSCOPE experiment. To improve the accuracy of measurements, the STEP mission would use up to four pairs of cylinders instead of just one pair. In addition, a tank of liquid helium would keep the temperature of the cylinders constant. The shell masses would also be surrounded with a superconducting shell. According to theoretical preliminary estimates, the STEP mission would allow to confirm or to disconfirm the Equivalence Principle with accuracy $|m_i - m_g|/m_i \sim 10^{-18}$. The peculiarity of the GG space mission is that the satellite would quickly spin about its main axis, which provides the accuracy of the measurements of about $|m_i - m_g|/m_i \sim 10^{-17}$.

It is easy to understand that a violation of the weak (i.e. Galileo's) Equivalence Principle means that Einstein's Equivalence Principle is also violated, but the opposite statement is not true. In this book, we consider such unexpected special situations, where Einstein's Equivalence Principle or weak Equivalence Principle are violated from theoretical point of view and briefly discuss the possible experimental methods to detect such violations.

In Chapter 1, written by Prof. A.G. Lebed (University of Arizona, USA), the unexpected breakdowns of Einstein's Equivalence Principle for composite quantum bodies are considered. The author points out that all bodies in General Relativity have a quantum nature, which, thus, has to be explicitly theoretically taken into account both for passive and active gravitational masses. The simplest composite quantum body, a hydrogen atom, is considered and the corresponding passive and active gravitational quantum mass operators are calculated, including contributions from electron kinetic energy, K, and from its potential Coulomb energy, P. It is shown that the above-mentioned operators do not commute with the energy operator in the absence of gravitational field due to the fact that the gravitational field is not only coupled to the total energy, $E = K + P$, but also to the so-called virial term, $V = 2K + P$. As a result, it is shown that Einstein's Equivalence Principle is broken both at microscopic and macroscopic levels for quantum bodies. Suppose the energy of a quantum state in a hydrogen atom is fixed, $E = E_1$ (i.e. the atom is in its ground state), then, as shown, quantum measurement of the passive gravitational mass with small probabilities would give the following quantized values: $m_g^n = E_n/c^2$, where E_n are energies of the excited electron levels. This always breaks Einstein's Equivalence Principle at a microscopic level. As for the macroscopic level, most macroscopic ensembles of atoms are shown to respect Einstein's Equivalence Principle. Nevertheless, it is also shown that for some special quantum macroscopic ensembles — ensembles of the coherent superpositions of the stationary quantum states (which we call Gravitational demons) — Einstein's Equivalence Principle between passive gravitational mass and energy is broken. We show that for such superpositions, the average values of passive gravitational masses are not related to the average values of energy by the famous Einstein equation, i.e. $m_g \neq E/c^2$. Possible experiments for Earth-bound laboratories are briefly discussed and contrasted with the existing space missions in terms of the potential to discover the possible breakdown of the Equivalence Principle.

In Chapter 2, Profs. D. Singleton and J. Contreras (California State University, Fresno, USA) and Prof. E. T. Akhmedov (Moscow Institute of Physics and Technology, Russia) show that there are certain quantum mechanical phenomena which are able to tell the difference between a gravitational field and a system which is accelerating. In particular, the two phenomena which they study are Hawking radiation seen by an observer around a non-rotating black hole versus Unruh radiation seen by an observer accelerating uniformly. They find that both observers see radiation, but there are differences in the details of this radiation which allow the observers to say if they are in the gravitational field of a black hole or if they are in empty space, accelerating uniformly. They study the details of the radiation of these two cases using an Unruh–DeWitt detector. The Unruh–DeWitt detector is a quantum two-state system where the extent to which the upper state is populated will tell one if there is radiation associated with a particular space-time and the details of this radiation. The source of this breaking of the Equivalence Principle rests with the fact that the Equivalence Principle is a local principle, while quantum effects like Hawking radiation and Unruh radiation are inherently non-local since they depend on the definition of the vacuum which requires the knowledge of states defined globally over the space-time. The authors also briefly discuss the response of the Unruh–DeWitt detector for the acceleration coming from circular motion versus uniform, linear acceleration. This comparison may allow for the experimental confirmation of Unruh radiation, using electrons in circular particle accelerators.

In Chapter 3, Prof. M. Gasperini (University of Bari, Italy) stresses that we are used to thinking, according to everyday experience, that the gravitational force is attractive. This is not always the case, however. There are physical situations, not necessarily of "exotic" type, where the effective gravitational interaction turns out to be repulsive. This is possible, as is well known, even in the context of the standard Einstein model of gravity, when — for instance — the source of gravity is a matter fluid with negative (and large enough) effective pressure. Another possible origin of repulsive gravitational forces is a local deviation from the Lorentz symmetry principle which, in the absence of gravity, is at the foundation of the theory of Special Relativity. But why should such a symmetry be broken? Since it is expected to hold exactly in vacuum and not, for instance, in the presence of a thermal bath, i.e. of a medium at constant non-vanishing temperature, filling the whole spacetime. In that case, the theory of gravity has to be modified to take into account the thermal effects, and the gravitational forces may become repulsive at high enough temperature. Also, the gravitational and inertial masses are differently affected by the external temperature so that we can expect, in general, possible violations of the Equivalence Principle. Finally, where should we expect that the repulsive gravitational forces produced by the temperature (and/or by a more general local breaking of the Lorentz symmetry) may have important physical consequences? There is an obvious answer to this question: in the primordial cosmic epochs, where the energy scale is very large, the temperature is very high, gravity is the dominant interaction, and the presence of repulsive forces can stabilize our primordial Universe, and possibly prevent the collapse of the space-time geometry into an (non-physical) initial singularity.

Chapter 4, written by Prof. J. Q. Quach (University of Adelaide, Australia) *et al.*, is devoted to the weak Equivalence Principle, which is seen as an unpinning assumption in the theory of General Relativity. It states that all laws of motion for freely falling particles are the same as in an unaccelerated reference frame. General Relativity treats free-falling particles as point particles, an assumption that falls apart at the quantum scale where a particle's wavefunction must be considered. Violations of the weak Equivalence Principle by quantum particles are seen as a natural gateway toward studying quantum gravity. Experimental observation of violations of this principle have thus far eluded physicists given their infinitesimal signatures. Here the authors utilize a unique insight that for the weak Equivalence Principle to hold, one should not be able to gain any information about a particle's mass from its interaction with gravitational fields. This motivates the use of information techniques to investigate the principle violation. Using this approach, they find that the weak Equivalence Principle holds for a quantum particle in a uniform gravitational field, but strikingly is violated in non-uniform and time-dependent gravitational fields, such as gravitational waves. At the nanoscale, the Casimir effect is a pressure that results between objects which comes from a background of electromagnetic waves that emanate from quantum vacuum fluctuations. Interestingly, it has been recently suggested that a gravitational analogue of the Casimir effect, from gravitational wave fluctuations, could exist between superconductors. Using this information-theoretic approach, the authors theoretically demonstrate that a signature of weak Equivalence Principle violation could manifest in the form of a detectable gravitational Casimir effect between superconductors.

In Chapter 5, Profs. S. Capozziello (University of Napoli, Italy) and G. Lambiase (University of Salerno, Italy) stress that the Equivalence Principle is one of the fundamental principles of Nature. It is based on the former observation by Galilei and Newton who stated that gravitational and inertial masses coincide for any massive bodies independent of their own constitution. Einstein extended such a statement pointing out that, thanks to this property, inertial effects are locally indistinguishable from gravitational effects, which means the equivalence between inertial and gravitational masses and, furthermore, any gravitational field can be locally canceled finding out a local frame where accelerations are null. Validity or violation of the Equivalence Principle represents one of the main challenges of modern physics, both from theoretical and experimental points of view. In this respect, high precision experiments are conceived and realized for testing the theory of General Relativity as well as any theory of gravity based on spacetime measurements and geodesic motions. The crucial point of testing the Equivalence Principle is showing if it holds both at classical and quantum levels. If violation is somehow proved, several important foundations of modern physics have to be revised. Here, the authors critically discuss the Einstein Equivalence Principle both in General Relativity and in some of its modifications and extensions considering, mainly, its foundation and its relation with Quantum Field Theory. The aim of the chapter is to present a comprehensive and critical picture of the problem.

This book is devoted to the 110th anniversary of the formulation of Einstein's Equivalence Principle. The editor dedicates his own contributions to the book to his wife, Natalia, for her loving support and patience.

A.G. Lebed

Contents

Chapter 1

Breakdown of the Einstein's Equivalence Principle for a Quantum Body

Andrei G. Lebed

Department of Physics, University of Arizona,
1118 E. 4th Street, Tucson, AZ 85721, USA
L.D. Landau Institute for Theoretical Physics, RAS, 4 Kosygina Street,
Moscow 117334, Russia

lebed@arizona.edu

We discuss our recent theoretical results about inequivalence between passive and active gravitational masses and energy in semiclassical variant of General Relativity, where gravitational field is not quantized but matter is quantized. In particular, we consider the simplest composite quantum body — a hydrogen atom. We concentrate our attention on the following physical effects, related to the electron mass. The first one is breakdown of the equivalence between passive gravitational mass and energy at a microscopic level. Indeed, as we show, the quantum measurement of gravitational mass can give result, which is different from the expected, $m^g \neq m_e + \frac{E_1}{c^2}$, where electron is initially in its ground state with $E = E_1$; m_e is the bare electron mass. The second effect is that the expectation values of both passive and active gravitational masses of stationary quantum states are equivalent to the expectation values of energy. The most spectacular effects are inequivalence of passive and active gravitational masses and energy at a macroscopic level for ensembles of the coherent superpositions of stationary quantum states (which we call Gravitational demons). We demonstrate that, for such Gravitational demons, the expectation values of passive and active gravitational masses are not related to the expectation values of energy by the famous Einstein's equation, $m^g \neq \frac{E}{c^2}$.

1. Introduction

The Galileo Galilei's Equivalence Principle between gravitational and inertial masses together with the local Lorentz invariance of space-time are called the Einstein's Equivalence Principle. It is known to be a cornerstone of the classical General Relativity.[1,2] Validity of this principle for ordinary matter has been established so far with great accuracy, $\frac{|m_i - m_g|}{m_i} \leq 10^{-16} - 10^{-17}$, in the recent space mission "MICROSCOPE" (see Refs. 3 and 4), where m_g and m_i are gravitational and inertial masses, respectively. In literature, there are widely discussed possible new space missions, "Galileo Galilei"[5] and "STEP",[6] which may increase the above-mentioned accuracy up to $\frac{|m_i - m_g|}{m_i} \sim 10^{-19}$.

The quantum theory of gravity has not been developed yet, but the numerous speculations on this topic predict that the Einstein's Equivalence Principle may be broken at extremely high energies, $E \sim 10^{28}$eV, which will never be accessible for our experimental studies. Nevertheless, recently we have shown (see Refs. 7–10) that even semiclassical variant of General Relativity, where the field is not quantized but the matter is quantized, predicts breakdown of the Einstein's Equivalence Principle at low enough experimentally accessible energies. In particular, we have shown[7,9] that the Einstein's Equivalence Principle is broken for passive gravitational mass at a microscopic level for a composite quantum body. Indeed, electron in the hydrogen atom with constant stationary energy, E_n, is not characterized by constant passive gravitational mass.[7,9] According to the above cited works, there exists a small (but non-zero) probability that the quantum measurement of the electron mass gives the value $m^g \neq m_e + \frac{E_n}{c^2}$. We have also shown[9,10] that, although for stationary quantum states the expectation values of passive gravitational mass are equivalent to the expectation values of energy, it is possible to create the so-called Gravitational demons, which break the above-mentioned equivalence at a macroscopic level. Gravitational demons are defined as the macroscopic ensembles of the coherent quantum superpositions of the stationary quantum states. As shown in Refs. 9 and 10, the expectation values of the passive gravitational mass and energy of the demons are not related to each other by the famous Einstein's equation, $E \neq m^g c^2$. The situation with active gravitational mass, as shown,[8,9] is also very interesting since it breaks the Einstein's Equivalence Principle and the above-mentioned equation at a macroscopic level for the introduced above Gravitational demons. Nevertheless, it is shown[9,10] that the Einstein's Equivalence Principle survives for the majority of the quantum ensembles, including ensembles of the stationary quantum states.

2. Goal

In Secs. 3 and 4, we show that in the simplest composite quantum body — a hydrogen atom — the Einstein's Equivalence Principle is always broken at a microscopic level.[7,9] We demonstrate that quantum measurement of gravitational mass, m^g, in quantum state with definite energy, E_1, can always give unexpected results, $m^g = m_e + E_n/c^2$,[7,9] where E_n is one of the possible electron energy levels in the hydrogen atom, m_e is the bare electron mass, c is the velocity of light. In Secs. 6 and 7, we discuss in detail the breakdown of the Einstein's Equivalence Principle between passive gravitational mass and energy (i.e. inertial mass) at a macroscopic level, which was first suggested by us in Refs. 9 and 10. We show that the accepted by majority of physicists accuracy of the validity of the Equivalence Principle, $\frac{|m_i-m_g|}{m_i} \leq 10^{-17}\text{--}10^{-16}$, is overestimated since they experimentally studied only usual condensed matter samples. Below, we discuss behavior of several different macroscopic ensembles of the hydrogen atoms. In agreement with the above-mentioned experiments, we show that the Einstein's Equivalence Principle is valid for almost all of them (see Sec. 5). We demonstrate that this equivalence survives for macroscopic ensembles of the stationary quantum electron states in a hydrogen atom due to the so-called quantum virial theorem.[11] On the other hand, we construct such quantum ensemble, as Gravitational demon (in analogy with Maxwell demon), which breaks, as discussed in Secs. 6 and 7, the Einstein's Equivalence Principle at a macroscopic level.[9,10]

Note that the virial term was first suggested in Ref. 12 (see also Refs. 13 and 14) for the classical model of a hydrogen atom. In particular, it was shown that an external gravitational field is coupled not to the total energy, $E = K+P$, but to the following combination: $E+V$, where the virial term, $V = 2K + P$, with K and P being kinetic and potential energies, respectively. Nevertheless, it was claimed[13,14] that the virial term disappears in classical General Relativity, if we choose the local proper coordinates in the gravitational field. Therefore, we suggest in this chapter two methods to calculate passive gravitational mass: one, using gravitational field as a perturbation in the Minkowski's metric[9] (see Secs. 4 and 6), and another one — using the local proper coordinates[9,10] (see Secs. 3 and 7). We show that both methods give the same results for the breakdown of the Einstein's Equivalence Principle at microscopic and macroscopic levels. In Secs. 4.4 and 7.1, some experimental aspects of the above-mentioned breakdowns of the Equivalence Principle are discussed. In particular, we pay attention that it is not necessary to conduct very expensive experiments in space. It is possible to create the Gravitational demons in the Earth's labora- tories and, thus, to discover the breakdown of the Einstein's Equivalence Principle. Although the experiments are expected to be rather difficult, the effect of the breakdown of the Equivalence Principle may be very large and, in principle, even may be of the order of unity. In Sec. 8, we show that the Gravitational demons also break the Einstein's Equivalence Principle between active gravitational mass and energy,[8,9] although for the majority of quantum states it survives.

3. Inequivalence Between Passive Gravitational Mass and Energy at a Microscopic Level (Method-1)

3.1. Electron wave function in a hydrogen atom with a definite energy in the absence of gravitational field

Let us suppose that, at $t < 0$, there is no gravitational field and electron is in its ground state in a hydrogen atom, which is characterized by the following wave function:

$$\Psi_1(r,t) = \exp\left(\frac{-im_ec^2t}{\hbar}\right)\exp\left(\frac{-iE_1t}{\hbar}\right)\Psi_1(r), \qquad (1)$$

which is a solution of the corresponding Schrödinger equation:

$$i\hbar\frac{\partial\Psi_1(r,t)}{\partial t} = \left[m_ec^2 - \frac{\hbar^2}{2m_e}\left(\frac{\partial^2}{\partial x^2} + \frac{\partial^2}{\partial y^2} + \frac{\partial^2}{\partial z^2}\right) - \frac{e^2}{r}\right]\Psi_1(r,t). \qquad (2)$$

(Above, E_1 is electron ground-state energy, r is distance between electron with coordinates (x,y,z) and proton; $\hbar$ is the Planck constant, c is the velocity of light.)

3.2. Electron wave functions in a hydrogen atom in the presence of gravitational field

We perform the following Gedanken experiment at $t = 0$. We switch on a weak centrosymmetric (e.g. the Earth's) gravitational field, with position of center of mass of the atom (i.e. proton) being fixed in the field by some forces of non-gravitational origin. In a weak field approximation, curved space-time is characterized by the following interval[1,2]:

$$ds^2 = -\left(1 + 2\frac{\phi}{c^2}\right)(cdt)^2 + \left(1 - 2\frac{\phi}{c^2}\right)(dx^2 + dy^2 + dz^2), \quad \phi = -\frac{GM}{R}. \quad (3)$$

Below we introduce the proper local coordinates,

$$\tilde{t} = \left(1 + \frac{\phi}{c^2}\right)t, \quad \tilde{x} = \left(1 - \frac{\phi}{c^2}\right)x,$$

$$\tilde{y} = \left(1 - \frac{\phi}{c^2}\right)y, \quad \tilde{z} = \left(1 - \frac{\phi}{c^2}\right)z, \quad (4)$$

where the interval has the standard Minkowski's form,[1,2]

$$d\tilde{s}^2 = -(cd\tilde{t})^2 + (d\tilde{x}^2 + d\tilde{y}^2 + d\tilde{z}^2). \quad (5)$$

(Here, we point out that, since we are interested in calculations of some quantum amplitudes with the first-order accuracy with respect to the small parameter $|\frac{\phi}{c^2}| \ll 1$, we disregard in Eqs. (3)–(5) and therein below all terms of the second order, $\frac{\phi^2}{c^4}$. We stress that near the Earth's surface the above-discussed parameter is small and is equal to $|\frac{\phi}{c^2}| \sim 10^{-9}$.)

Due to the property of the local Lorentz invariance of a space-time in General Relativity, new wave functions, expressed in the local proper coordinates (4) (with fixed proton position), satisfy at $t, \tilde{t} > 0$ the similar Schrödinger equation, if we disregard the tidal terms in the Hamiltonian:

$$i\hbar\frac{\partial \tilde{\Psi}(\tilde{r}, \tilde{t})}{\partial \tilde{t}} = \left[m_e c^2 - \frac{\hbar^2}{2m_e}\left(\frac{\partial^2}{\partial \tilde{x}^2} + \frac{\partial^2}{\partial \tilde{y}^2} + \frac{\partial^2}{\partial \tilde{z}^2}\right) - \frac{e^2}{\tilde{r}}\right]\tilde{\Psi}(\tilde{r}, \tilde{t}). \quad (6)$$

(We note that it is possible to show that the above-mentioned tidal terms have relative order of $\frac{r_0}{R_0}$, where r_0 is the Bohr radius and R_0 is distance between a hydrogen atom and center of source of gravitational field. Near the Earth's surface they are very small and are of the relative order of $\frac{r_0}{R_0} \sim 10^{-17}$).

It is very important that the wave function (1) is not a solution of the Schrödinger equation (6) anymore and, thus, is not characterized by definite energy and weight in the gravitational field (3). Moreover, we can express a general solution of Eq. (6) in the proper local coordinates in the following way:

$$\tilde{\Psi}(\tilde{r}, \tilde{t}) = \exp\left(\frac{-im_ec^2\tilde{t}}{\hbar}\right) \sum_{n=1}^{\infty} \tilde{a}_n \Psi_n(\tilde{r}) \exp\left(\frac{-iE_n\tilde{t}}{\hbar}\right), \tag{7}$$

where the wave functions $\Psi_n(\tilde{r})$ are solutions[11] for the so-called nS atomic orbits of a hydrogen atom with energies E_n and are normalized in the proper local space,

$$\int \Psi_n^2(\tilde{r}) \, d^3\tilde{r} = 1. \tag{8}$$

(Here, we stress that, as easy to show, only $1S \rightarrow nS$ quantum transitions amplitudes are non-zero in a hydrogen atom in the gravitational field (3), which correspond only to real wave functions. Thus, we keep in Eq. (7) only nS atomic orbits.)

Note that the normalized wave function (1) can be expressed in the proper local space-time coordinates (4) in the following way:

$$\Psi_1(\tilde{r}, \tilde{t}) = \exp\left[\frac{-im_ec^2(1 - \frac{\phi}{c^2})\tilde{t}}{\hbar}\right] \exp\left[\frac{-iE_1(1 - \frac{\phi}{c^2})\tilde{t}}{\hbar}\right]$$
$$\times \left(1 + \frac{\phi}{c^2}\right)^{3/2} \Psi_1\left[\left(1 + \frac{\phi}{c^2}\right)\tilde{r}\right], \tag{9}$$

We stress that the gravitational field (3) can be considered as a sudden perturbation to the Hamiltonian (2), therefore, the wave functions (7) and (9) have to be equal to each other at $t = \tilde{t} = 0$:

$$\left(1 + \frac{\phi}{c^2}\right)^{3/2} \Psi_1\left[\left(1 + \frac{\phi}{c^2}\right)\tilde{r}\right] = \sum_{n=1}^{\infty} \tilde{a}_n \Psi_n(\tilde{r}). \tag{10}$$

From (10), it directly follows that

$$\tilde{a}_1 = \left(1 + \frac{\phi}{c^2}\right)^{3/2} \int_0^{\infty} \Psi_1\left[\left(1 + \frac{\phi}{c^2}\right)\tilde{r}\right] \Psi_1(\tilde{r}) \, d^3\tilde{r} \tag{11}$$

and

$$\tilde{a}_n = \left(1 + \frac{\phi}{c^2}\right)^{3/2} \int_0^{\infty} \Psi_1\left[\left(1 + \frac{\phi}{c^2}\right)\tilde{r}\right] \Psi_n(\tilde{r}) \, d^3\tilde{r}, \quad n > 1. \tag{12}$$

3.3. *Probabilities and amplitudes*

In this section, we calculate quantum mechanical amplitudes (11) and (12) in a linear approximation with respect to the gravitational potential,

$$\tilde{a}_1 \approx 1, \tag{13}$$

and

$$\tilde{a}_n = \left(\frac{\phi}{c^2} \right) \int_0^\infty \left[\frac{d\Psi_1(\tilde{r})}{d\tilde{r}} \right] \tilde{r}\Psi_n(\tilde{r})d^3\tilde{r}, \quad n > 1. \tag{14}$$

We point out that the wave function (7) is a series of wave functions, which have definite electron weights in the gravitational field (3). It is easy to understand that they are characterized by the following definite electron passive gravitational masses,

$$m_n^g = m_e + \frac{E_n}{c^2}. \tag{15}$$

In accordance with the general properties of quantum mechanics, if we do a measurement of gravitational mass for wave function (1) and (9), we obtain quantized values (15) with the following probabilities: $\tilde{P}_n = |\tilde{a}_n|^2$, where $\tilde{a}_n$ are given by Eqs. (13) and (14).

Below, we show that in Eq. (14)

$$\int_0^\infty \left[\frac{d\Psi_1(\tilde{r})}{d\tilde{r}} \right] \tilde{r}\Psi_n(\tilde{r})d^3\tilde{r} = \frac{V_{1n}}{E_n - E_1}, \quad n > 1, \tag{16}$$

where $\hat{V}(\tilde{r})$ is the so-called quantum virial operator[11]:

$$\hat{V}(r) = -2\frac{\hbar^2}{2m_e} \left(\frac{\partial^2}{\partial\tilde{x}^2} + \frac{\partial^2}{\partial\tilde{y}^2} + \frac{\partial^2}{\partial\tilde{z}^2} \right) - \frac{e^2}{\tilde{r}}, \tag{17}$$

and

$$V_{1n} = \int_0^\infty \Psi_1(\tilde{r})\hat{V}(\tilde{r})\Psi_n(\tilde{r})d^3\tilde{r}. \tag{18}$$

To this end, we rewrite the obtained above Schrödinger equation (6) in the gravitational field (3), in terms of the initial coordinates (x, y, z):

$$(m_e c^2 + E_1)\Psi_1 \left[\left(1 - \frac{\phi}{c^2} \right) r \right] = \left[m_e c^2 - \frac{1}{(1 - \phi/c^2)^2} \frac{\hbar^2}{2m} \right.$$
$$\times \left(\frac{\partial^2}{\partial x^2} + \frac{\partial^2}{\partial y^2} + \frac{\partial^2}{\partial z^2} \right) - \frac{1}{(1 - \phi/c^2)} \frac{e^2}{r} \right]$$
$$\times \Psi_1 \left[\left(1 - \frac{\phi}{c^2} \right) r \right]. \tag{19}$$

Then, taking into account only terms of the first order with respect to the small parameter $\left|\frac{\phi}{c^2}\right| \ll 1$, we obtain

$$E_1\Psi_1(r) - \frac{\phi}{c^2}E_1 r\left[\frac{d\Psi_1(r)}{dr}\right] = \left[-\frac{\hbar^2}{2m_e}\left(\frac{\partial^2}{\partial x^2} + \frac{\partial^2}{\partial y^2} + \frac{\partial^2}{\partial z^2}\right) - \frac{e^2}{r}\right.$$
$$\left. + \frac{\phi}{c^2}\hat{V}(r)\right]\left\{\Psi_1(r) - \frac{\phi}{c^2}r\left[\frac{d\Psi_1(r)}{dr}\right]\right\} \quad (20)$$

and

$$-E_1 r\left[\frac{d\Psi_1(r)}{dr}\right] = \left[-\frac{\hbar^2}{2m_e}\left(\frac{\partial^2}{\partial x^2} + \frac{\partial^2}{\partial y^2} + \frac{\partial^2}{\partial z^2}\right) - \frac{e^2}{r}\right]$$
$$\times\left[-r\frac{d\Psi_1(r)}{dr}\right] + \hat{V}(r)\Psi_1(r). \quad (21)$$

Here, we multiply Eq. (21) on $\Psi_n(r)$ and integrate over space,

$$-E_1\int_0^\infty \Psi_n(r)r\left[\frac{d\Psi_1(r)}{dr}\right]d^3r$$
$$= \int_0^\infty \Psi_n(r)\left[-\frac{\hbar^2}{2m_e}\left(\frac{\partial^2}{\partial x^2} + \frac{\partial^2}{\partial y^2} + \frac{\partial^2}{\partial z^2}\right) - \frac{e^2}{r}\right]$$
$$\times\left[-r\frac{d\Psi_1(r)}{dr}\right]d^3r + \int_0^\infty \Psi_n(r)\hat{V}(r)\Psi_1(r)d^3r. \quad (22)$$

Taking account of the fact that the Hamiltonian operator is the Hermitian one, we rewrite Eq. (22) as

$$E_1\int_0^\infty \Psi_n(r)r\left[\frac{d\Psi_1(r)}{dr}\right]d^3r = E_n\int_0^\infty \Psi_n(r)r\left[\frac{d\Psi_1(r)}{dr}\right]d^3r$$
$$- \int_0^\infty \Psi_1(r)\hat{V}(r)\Psi_n(r)d^3r. \quad (23)$$

Now, Eqs. (16)–(18) directly follow from Eq. (23).

As a result of our calculations, the amplitudes (14) and the corresponding probabilities for $n > 1$ can be rewritten as functions of matrix elements (18) of the quantum virial operator (17),

$$\tilde{a}_n = \left(\frac{\phi}{c^2}\right)\frac{V_{1n}}{E_n - E_1} \quad (24)$$

and, therefore,

$$\tilde{P}_n = |\tilde{a}_n|^2 = \left(\frac{\phi}{c^2}\right)^2\left(\frac{V_{1n}}{E_n - E_1}\right)^2. \quad (25)$$

Note that near the Earth's surface, where $\frac{\phi^2}{c^4} \approx 0.49 \times 10^{-18}$, the probability for $n = 2$ in a hydrogen atom can be evaluated as

$$\tilde{P}_2 = |\tilde{a}_2|^2 = 1.5 \times 10^{-19}, \tag{26}$$

where

$$\frac{V_{12}}{E_2 - E_1} = 0.56. \tag{27}$$

We stress that non-zero matrix elements (18) of the virial operator (17) for $n \neq 1$ are also responsible for breakdown of the equivalence between active gravitational mass and energy for a quantum body with internal degrees of freedom.[8,9]

4. Inequivalence Between Passive Gravitational Mass and Energy at a Microscopic Level (Method-2)

4.1. *Schrödinger equation with a definite energy in the absence of gravitational field*

As in Sec. 3, gravitational field is supposed to be zero at $t < 0$ and electron occupies ground state in a hydrogen atom at $t < 0$ with the wave function (1). As we already discussed, the wave function (1) corresponds to the $1S$ electron orbit and is known to be a ground-state solution of the Schödinger equation (2).

4.2. *Schrödinger equation in the presence of gravitational field*

Let us consider the same Gedanken experiment as in Sec. 3. We switch on the weak gravitational field (3) at $t = 0$ and obtain Eq. (6) for the wave functions in the proper local space-time coordinates (4). The difference is that, in this section, we rewrite Eq. (6) in the initial space-time coordinates, (t, x, y, z),

$$
\begin{aligned}
i\hbar \frac{\partial \Psi(\mathbf{r}, t)}{\partial t} = \Bigg\{ &\left[m_e c^2 - \frac{\hbar^2}{2m_e} \left(\frac{\partial^2}{\partial x^2} + \frac{\partial^2}{\partial y^2} + \frac{\partial^2}{\partial z^2} \right) - \frac{e^2}{r} \right] \\
&+ \left(\frac{\phi}{c^2} \right) \left[m_e c^2 - \frac{\hbar^2}{2m_e} \left(\frac{\partial^2}{\partial x^2} + \frac{\partial^2}{\partial y^2} + \frac{\partial^2}{\partial z^2} \right) \right. \\
&\left. - \frac{e^2}{r} + \hat{V}(\mathbf{r}) \right] \Bigg\} \Psi(\mathbf{r}, t),
\end{aligned} \tag{28}
$$

where the quantum virial operator,[11] $\hat{V}(r)$, is equal to (17). It directly follows from Eq. (28) that the external gravitational field (3) is coupled not only to Hamiltonian (2) but also to the virial operator (17). We pay attention to the fact that the quantum virial term (17) does not commute with the Hamiltonian (2), therefore, it breaks at a microscopic level the equivalence between the passive gravitational mass and energy for electron in a hydrogen atom.

4.2.1. More general Lagrangian

In this section, we derive Hamiltonian (28) from the more general Lagrangian. Below, we consider the Lagrangian of the following three-body system: a hydrogen atom and the Earth in inertial coordinate system, where we treat the gravitational field (3) as a small perturbation in the Minkowski's space-time. In this case, we can make use of the results of Ref. 12, where the corresponding n-body Lagrangian is calculated:

$$L = L_{kin} + L_{em} + L_G + L_{e,G}, \qquad (29)$$

where L_{kin}, L_{em}, L_G, and $L_{e,G}$ are kinetic, electromagnetic, gravitational and electric-gravitational parts of the Lagrangian, correspondingly. We recall that, in the accepted approximation, we keep in the Lagrangian and Hamiltonian only terms of the order of $(v/c)^2$ and $|\phi|/c^2$ as well as keep only classical kinetic and the Coulomb electrostatic potential energies couplings to external gravitational field. It is easy to show that, in our case, different parts of the Lagrangian (29) can be simplified:

$$L_{kin} + L_{em} = -Mc^2 - m_p c^2 - m_e c^2 + m_e \frac{\mathbf{v}^2}{2} + \frac{e^2}{r}, \qquad (30)$$

$$L_G = G\frac{m_p M}{R} + G\frac{m_e M}{R} + \frac{3}{2}G\frac{m_e M}{R}\frac{\mathbf{v}^2}{c^2}, \qquad (31)$$

$$L_{e,G} = -2G\frac{M}{Rc^2}\frac{e^2}{r}, \qquad (32)$$

where we use the inequality $m_p \gg m_e$, with m_p being the bare proton mass.

If we keep only terms in the Lagrangian related to electron motion (as usual, proton is supposed to be supported by some non-gravitational forces), then we can write the Lagrangian (30)–(32) in the following simple form:

$$L = -m_e c^2 + m_e \frac{\mathbf{v}^2}{2} + \frac{e^2}{r} - \frac{\phi(R)}{c^2} \left[m_e c^2 + 3 m_e \frac{\mathbf{v}^2}{2} - 2 \frac{e^2}{r} \right],$$

$$\phi(R) = -G \frac{M}{R}. \tag{33}$$

It is possible to show that the corresponding quantum electron Hamiltonian is

$$\hat{H} = \left\{ \left[m_e c^2 - \frac{\hbar^2}{2m_e} \left(\frac{\partial^2}{\partial x^2} + \frac{\partial^2}{\partial y^2} + \frac{\partial^2}{\partial z^2} \right) - \frac{e^2}{r} \right] \right.$$
$$\left. + \left(\frac{\phi}{c^2} \right) \left[m_e c^2 - \frac{\hbar^2}{2m_e} \left(\frac{\partial^2}{\partial x^2} + \frac{\partial^2}{\partial y^2} + \frac{\partial^2}{\partial z^2} \right) - \frac{e^2}{r} + \hat{V}(\mathbf{r}) \right] \right\}. \tag{34}$$

(We note that Eq. (34) exactly coincides with the electron Hamiltonian (28), obtained above.)

4.2.2. *More general Hamiltonian*

Let us derive the electron Hamiltonian (28),(34) from more general arguments. The so-called Stark effect (i.e. the mixing effect between even and odd wave functions in a hydrogen atom in gravitational field) was carefully studied in Ref. 16 in the weak external gravitational field (3). Note that the corresponding electron Hamiltonian was derived in $1/c^2$ approximation and a possibility of center of mass of the atom motion was taken into account. The main peculiarity of the above-mentioned calculations was the fact that not only terms of the order of ϕ/c^2 were calculated, as in our case, but also terms of the order of ϕ'/c^2, where we use a symbolic notation ϕ' for the first derivatives of gravitational potential. In accordance with the existing tradition, we refer to the latter terms as to the tidal ones. It is important that the Hamiltonian (3.24) was obtained in Ref. 16 directly from the Dirac equation in a curved space-time of General Relativity. As shown in Ref. 16, it can be rewritten for the corresponding Schrödinger equation as a sum of the four terms:

$$\hat{H}(\hat{\mathbf{P}}, \hat{\mathbf{p}}, \tilde{\mathbf{R}}, \mathbf{r}) = \hat{H}_0(\hat{\mathbf{P}}, \hat{\mathbf{p}}, r) + \hat{H}_1(\hat{\mathbf{P}}, \hat{\mathbf{p}}, \tilde{\mathbf{R}}, r) + \hat{H}_2(\hat{\mathbf{p}}, \mathbf{r}) + \hat{H}_3(\hat{\mathbf{P}}, \hat{\mathbf{p}}, \tilde{\mathbf{R}}, \mathbf{r}), \tag{35}$$

$$\hat{H}_0(\hat{\mathbf{P}}, \hat{\mathbf{p}}, r) = m_e c^2 + m_p c^2 + \left[\frac{\hat{\mathbf{P}}^2}{2(m_e + m_p)} + \frac{\hat{\mathbf{p}}^2}{2\mu} \right] - \frac{e^2}{r}, \tag{36}$$

$$\hat{H}_1(\hat{\mathbf{P}}, \hat{\mathbf{p}}, \tilde{\mathbf{R}}, r) = \left\{ m_e c^2 + m_p c^2 + \left[3\frac{\hat{\mathbf{P}}^2}{2(m_e + m_p)} + 3\frac{\hat{\mathbf{p}}^2}{2\mu} - 2\frac{e^2}{r} \right] \right\}$$

$$\times \left(\frac{\phi - \mathbf{g}\tilde{\mathbf{R}}}{c^2} \right), \tag{37}$$

$$\hat{H}_2(\hat{\mathbf{p}}, \mathbf{r}) = \frac{1}{c^2}\left(\frac{1}{m_e} - \frac{1}{m_p} \right) [-(\mathbf{gr})\hat{\mathbf{p}}^2 + i\hbar \mathbf{g}\hat{\mathbf{p}}]$$

$$+ \frac{1}{c^2}\mathbf{g}\left(\frac{\hat{\mathbf{s}}_e}{m_e} - \frac{\hat{\mathbf{s}}_p}{m_p} \right)\hat{\mathbf{p}} + \frac{e^2(m_p - m_e)}{2(m_e + m_p)c^2}\frac{\mathbf{gr}}{r}, \tag{38}$$

$$\hat{H}_3(\hat{\mathbf{P}}, \hat{\mathbf{p}}, \tilde{\mathbf{R}}, \mathbf{r}) = \frac{3}{2}\frac{i\hbar \mathbf{g}\hat{\mathbf{P}}}{(m_e + m_p)c^2} + \frac{3}{2}\frac{\mathbf{g}(\hat{\mathbf{s}}_e + \hat{\mathbf{s}}_p) \times \hat{\mathbf{P}}}{(m_e + m_p)c^2}$$

$$- \frac{(\mathbf{gr})(\hat{\mathbf{P}}\hat{\mathbf{p}}) + (\hat{\mathbf{P}}\mathbf{r})(\mathbf{g}\hat{\mathbf{p}}) - i\hbar \mathbf{g}\hat{\mathbf{P}}}{(m_e + m_p)c^2}, \tag{39}$$

where $\mathbf{g} = -G\frac{M}{R^3}\mathbf{R}$. We use the following notations in Eqs. (35)–(39): $\tilde{\mathbf{R}}$ and $\hat{\mathbf{P}}$ stand for coordinate and momentum operator of a hydrogen atom center of mass, respectively, whereas $\mathbf{r}$ and $\hat{\mathbf{p}}$ stand for relative electron coordinate and electron momentum operator in center of mass coordinate system; $\mu = m_e m_p/(m_e + m_p)$. We point out that $\hat{H}_0(\hat{\mathbf{P}}, \hat{\mathbf{p}}, r)$ is the Hamiltonian of a hydrogen atom in the absence of the gravitational field. It is important that the Hamiltonian $\hat{H}_1(\hat{\mathbf{P}}, \hat{\mathbf{p}}, \tilde{\mathbf{R}}, r)$ describes couplings not only of the bare electron and proton masses to the gravitational field (3) but also couplings of electron kinetic and potential energies to the field. And finally, the Hamiltonians $\hat{H}_2(\hat{\mathbf{p}}, \mathbf{r})$ and $\hat{H}_3(\hat{\mathbf{P}}, \hat{\mathbf{p}}, \tilde{\mathbf{R}}, \mathbf{r})$ describe only the tidal effects.

Here, we strictly derive the Hamiltonian (28),(34), which has already been semi-quantitatively derived above, from the general Hamiltonian (35)–(39). As was mentioned, we use the approximation $m_p \gg m_e$ and, therefore, $\mu = m_e$. In particular, this allows us to consider proton as a heavy classical particle. We recall that we need to derive the Hamiltonian of the atom, whose proton is at rest with respect to the Earth. Therefore, we omit center of mass kinetic energy and center of mass momentum. As a result, the first two contributions to electron part of the total Hamiltonian (35)–(39) can be written in the following familiar form:

$$\hat{H}_0(\hat{\mathbf{p}}, r) = m_e c^2 + \frac{\hat{\mathbf{p}}^2}{2m_e} - \frac{e^2}{r} \tag{40}$$

and

$$\hat{H}_1(\hat{\mathbf{p}}, r) = \left\{ m_e c^2 + \left[3\frac{\hat{\mathbf{p}}^2}{2m_e} - 2\frac{e^2}{r} \right] \right\} \left(\frac{\phi}{c^2} \right), \tag{41}$$

where we place center of mass of the atom at point $\tilde{\mathbf{R}} = 0$. Now, let us study the first tidal term (38). At first, we pay attention that $|\mathbf{g}| \simeq |\phi|/R_0$. Then, as well known, in a hydrogen atom $|\mathbf{r}| \sim \hbar/|\mathbf{p}| \sim r_B$ and $\mathbf{p}^2/(2m_e) \sim e^2/r_B$. These values allow us to evaluate the first tidal term (38) in the Hamiltonian (35) as $H_2 \sim (r_B/R_0)(|\phi|/c^2)(e^2/r_B) \sim 10^{-17}(|\phi|/c^2)(e^2/r_B)$. Note that this value is 10^{17} times smaller than $H_1 \sim (|\phi|/c^2)(e^2/r_B)$. Therefore, we can disregard the contribution (38) to the total Hamiltonian (35). As to the second tidal term (39), we pay attention that it is exactly zero in the case, where $\mathbf{P} = 0$, considered in this chapter. Therefore, we can conclude that the derived electron Hamiltonian (40),(41) exactly coincides semi-quantitatively with (28),(34).

4.3. *Gravitational field as a perturbation to the Hamiltonian*

It is easy to understand that the gravitational field (3), under the condition of our Gedanken experiment, can be considered as the following sudden perturbation, $\hat{U}_1(\mathbf{r}, t)$, to the Hamiltonian (2):

$$\hat{U}_1(\mathbf{r}, t) = \left(\frac{\phi}{c^2} \right) \left[m_e c^2 - \frac{\hbar^2}{2m_e} \left(\frac{\partial^2}{\partial x^2} + \frac{\partial^2}{\partial y^2} + \frac{\partial^2}{\partial z^2} \right) - \frac{e^2}{r} + \hat{V}(\mathbf{r}) \right] \Theta(t), \tag{42}$$

where $\Theta(t)$ is the step function. In this case, a general solution of Eq. (28) can be written in the following way:

$$\Psi(r, t) = \exp\left(\frac{-i\tilde{m}_e c^2 t}{\hbar} \right) \Psi_1^1(r) \exp\left(\frac{-i\tilde{E}_1 t}{\hbar} \right)$$

$$+ \exp\left(\frac{-im_e c^2 t}{\hbar} \right) \sum_{n>1}^{\infty} a_n \Psi_n(r) \exp\left(\frac{-iE_n t}{\hbar} \right), \tag{43}$$

where the wave functions $\Psi_n(r)$ are the normalized solutions for the nS electron orbits in a hydrogen atom,

$$\int [\Psi_1^1(r)]^2 \, d^3r = 1; \quad \int [\Psi_n(r)]^2 \, d^3r = 1, \quad n > 1. \tag{44}$$

(Note that perturbation (42) results only in non-zero quantum transitions between $1S$ and nS electron orbits, thus, we keep in Eq. (43) only $\Psi_n(r)$ wave functions, which are real.)

In accordance with the standard time-dependent perturbation theory,[11] the corrected wave function of ground state, $\Psi_1^1(r)$, as well as the corrections to mass and energy of ground state can be written as

$$\Psi_1^1(r) = \Psi_1(r) + \left(\frac{\phi}{c^2}\right)\sum_{n>1}^{\infty}\frac{V_{n1}}{E_1 - E_n}\Psi_n(r),$$

$$\tilde{m}_e = \left(1 + \frac{\phi}{c^2}\right)m_e, \quad \tilde{E}_1 = \left(1 + \frac{\phi}{c^2}\right)E_1, \tag{45}$$

where V_{n1} is matrix element of the quantum virial operator (17):

$$V_{n1} = \int \Psi_n(r)\left[-2\frac{\hbar^2}{2m}\left(\frac{\partial^2}{\partial x^2} + \frac{\partial^2}{\partial y^2} + \frac{\partial^2}{\partial z^2}\right) - \frac{e^2}{r}\right]\Psi_1(r)d^3\mathbf{r}. \tag{46}$$

We note that the last term in Eq. (45) corresponds to the redshift in gravitational field. It is due to the expected contribution to passive gravitational mass from electron binding energy in the hydrogen atom. As to the coefficients a_n with $n \neq 1$ in Eq. (43), they can be also expressed in terms of the quantum virial operator matrix elements,

$$a_n = \left(\frac{\phi}{c^2}\right)\left(\frac{V_{n,1}}{E_n - E_1}\right), \quad n > 1, \tag{47}$$

and coincides with Eq. (24). Note that the wave function (43)–(47), corresponding to electron ground energy level in the presence of the gravitational field (3), is a series of eigenfunctions of electron energy operator, taken in the absence of the field. Therefore, if we measure energy, in electron quantum state (43)–(47), we obtain the following quantized values for electron gravitational mass:

$$m_n^g = m_e + \frac{E_n}{c^2}, \tag{48}$$

where we omit the redshift effect. We can state on the basis of Eqs. (43)–(48) that the expected Einstein's equation, $m = m_e + \frac{E_1}{c^2}$, survives in our case with probability close to 1, whereas with the following small probabilities,

$$P_n = |a_n|^2 = \left(\frac{\phi}{c^2}\right)^2\frac{V_{n1}^2}{(E_n - E_1)^2}, \quad n \neq 1, \tag{49}$$

it is broken. The reason for this breakdown is that, the virial term (17) does not commute with the Hamiltonian (2) in the absence of gravitational field. As a result, electron wave functions with definite passive gravitational masses are not characterized by definite energies in the absence of gravitational field. It is important that our current results coincide with that obtained in Sec. 3 by the different method.

4.4. *Some experimental aspects*

In this section, we describe another Gedanken experiment, where gravitational field is adiabatically switched on. In particular, we consider wave function (1) to be valid at $t \to -\infty$ and apply the following perturbation, due to the gravitational field (3), for the Hamiltonian (2):

$$\hat{U}_2(\mathbf{r}, t) = \left(\frac{\phi}{c^2}\right)\left[m_e c^2 - \frac{\hbar^2}{2m_e}\left(\frac{\partial^2}{\partial x^2} + \frac{\partial^2}{\partial y^2} + \frac{\partial^2}{\partial z^2}\right) - \frac{e^2}{r} + \hat{V}(\mathbf{r})\right]$$
$$\times \exp(\lambda t), \quad \lambda \to 0. \tag{50}$$

Then, at $t \simeq 0$ (i.e. in the presence of the field), the electron wave function can be expressed as

$$\Psi(r, t) = \exp\left(\frac{-i\tilde{m}_e c^2 t}{\hbar}\right)\Psi_1^1(r)\exp\left(\frac{-i\tilde{E}_1 t}{\hbar}\right)$$
$$+ \exp\left(\frac{-i m_e c^2 t}{\hbar}\right)\sum_{n>1}^{\infty} a_n\Psi_n(r)\exp\left(\frac{-iE_n t}{\hbar}\right). \tag{51}$$

In the case of adiabatic switching of gravitational field, application of the standard time-dependent perturbation theory[11] results in:

$$\Psi_1^1(r) = \Psi_1(r) + \left(\frac{\phi}{c^2}\right)\sum_{n>1}^{\infty}\frac{V_{n,1}}{E_1 - E_n}\Psi_n(r),$$
$$\tilde{m}_e = \left(1 + \frac{\phi}{c^2}\right)m_e, \quad \tilde{E}_1 = \left(1 + \frac{\phi}{c^2}\right)E_1, \tag{52}$$

and

$$a_n = 0, \quad P_n = 0, \quad n > 1. \tag{53}$$

Thus, in adiabatic limit, the phenomenon of quantization of passive gravitational mass (15),(48) disappears. This means that the possible experimental observation of the above-mentioned phenomenon has to be done in quickly changing gravitational field. It is important that step-like function, $\Theta(t)$, which was used to derive Eq. (48), does not mean motion of a hydrogen atom or a source of gravity with velocity higher than the speed of light. We can use step-like function if significant change of gravitational field happens quicker than the characteristic period of quasi-classical rotation of electron in a hydrogen atom. In the case under consideration, we need the time of the order of $\delta t \sim t_0 = \frac{2\pi\hbar}{E_2 - E_1} \sim 10^{-15}$s. We hope that there exist much more convenient quantum systems with higher values of the parameter t_0, where the above-discussed phenomenon could be observed. We recall that all excited energy levels are quasi-stationary and, thus, decay with time by emitting photons. Therefore, it is much more efficient to detect emitted photons than to directly measure a weight. (We also would like to stress that we did not take into account terms of the order of ϕ^2/c^4 in the gravitational field (3). Therefore, we did not calculated a weight of the atom with the accuracy up to ϕ^2/c^4.) As to the relatively small probabilities (25) of the mass quantization, they are not too small and can be compensated by large value of the Avogadro number, $N_A \approx 6 \times 10^{23}$. In other words, for macroscopic number of the atoms, we may have large number of emitted photons. For instance, the number of excited electrons (i.e. emitted photons) for 1000 moles of the atoms is estimated as

$$N_n = 2.95 \times 10^8 \times \left(\frac{V_{n,1}}{E_n - E_1} \right)^2, \quad N_2 = 0.9 \times 10^8. \tag{54}$$

Here, we would like to discuss more the status of the inequivalence between the passive gravitational mass and energy at a microscopic level, considered above. As we wrote, we did not show that the expectation values of the mass and energy were different on the scale of $m_e c^2 (\phi/c^2)^2$, since we did not considered the corresponding term in Eq. (3). But quantum mechanics, as well known, is not science about averages. We have calculated the probabilities (25) that quantum measurements will give the quantized results (15).

5. Einstein's Equivalence Principle for the Stationary Quantum States at a Macroscopic Level

As we stressed before, in the local proper space-time coordinates (4), the Schrödinger equation for the electron wave functions in a hydrogen atom can be approximately expressed in the standard form:

$$i\hbar \frac{\partial \Psi(\tilde{\mathbf{r}}, \tilde{t})}{\partial \tilde{t}} = \hat{H}(\hat{\tilde{p}}, \tilde{\mathbf{r}}) \Psi(\tilde{\mathbf{r}}, \tilde{t}), \tag{55}$$

with $\hat{H}(\hat{\tilde{p}}, \tilde{\mathbf{r}})$ being the standard Hamiltonian operator for a hydrogen atom. It is important to recall that, in Eq. (4) and below, we disregard all the tidal effects. In other words, we consider the atom as a point-like body and do not differentiate the gravitational potential with respect to the relative electron coordinates, $\mathbf{r}$ and $\tilde{\mathbf{r}}$. As we have already demonstrated, the disregarded tidal terms in the electron Hamiltonian (55) are very small and are of the relative order of $(r_B/R_0)|\phi/c^2| \sim 10^{-17}|\phi/c^2|$ in the Earth's gravitational field.

5.1. *Non-relativistic hydrogen atom*

First, let us consider the most important case, where we take account only of the kinetic and Coulomb potential energies in the non-relativistic Schrödinger equation for electron wave functions in a hydrogen atom:

$$i\hbar\frac{\partial\Psi(\tilde{\mathbf{r}},\tilde{t})}{\partial\tilde{t}} = \hat{H}_0(\hat{\tilde{p}},\tilde{\mathbf{r}})\Psi(\tilde{\mathbf{r}},\tilde{t}), \quad \hat{H}_0(\hat{\tilde{p}},\tilde{\mathbf{r}}) = m_e c^2 + \frac{\hat{\tilde{p}}^2}{2m_e} - \frac{e^2}{\tilde{r}}. \quad (56)$$

(In Eq. (56), as usual, e is the electron charge, $\tilde{r}$ is a distance between electron and proton, and $\hat{\tilde{p}} = -i\hbar\partial/\partial\tilde{\mathbf{r}}$ is electron momentum operator in the local proper coordinates.) Here, we consider inertial coordinate system, associated with the space-time coordinates (t, x, y, z) in Eq. (4) and treat the weak gravitational field (3) as a perturbation. As a result, we obtain the following Hamiltonian[7,9] (see also Eq. (28)).

$$\hat{H}_0(\hat{\mathbf{p}},\mathbf{r}) = m_e c^2 + \frac{\hat{\mathbf{p}}^2}{2m_e} - \frac{e^2}{r} + m_e\phi + \left(3\frac{\hat{\mathbf{p}}^2}{2m_e} - 2\frac{e^2}{r}\right)\frac{\phi}{c^2}. \quad (57)$$

Note that we can rewrite the Hamiltonian equation (57) in the following more convenient form:

$$\hat{H}_0(\hat{\mathbf{p}},\mathbf{r}) = m_e c^2 + \frac{\hat{\mathbf{p}}^2}{2m_e} - \frac{e^2}{r} + \hat{m}_e^g\phi, \quad (58)$$

where the passive gravitational mass operator of electron, $\hat{m}_e^g$, is introduced by the equation:

$$\hat{m}_e^g = m_e + \left(\frac{\hat{\mathbf{p}}^2}{2m_e} - \frac{e^2}{r}\right)\bigg/c^2 + \left(2\frac{\hat{\mathbf{p}}^2}{2m_e} - \frac{e^2}{r}\right)\bigg/c^2, \quad (59)$$

which has physical meaning of electron weight operator in the weak field (3). It is important that, in Eq. (59), only the first term corresponds to the bare electron mass, m_e. There exist two corrections to the bare electron mass: the expected second term, which corresponds to the electron binding energy contribution, and the non-trivial third term, which is the quantum virial contribution to the passive gravitational mass operator. As shown by us in Ref. 9 (see also Sec. 4), Eqs. (58) and (59) can be directly obtained from the Dirac equation in a weekly curved space-time (3), if we disregard all tidal terms.

Below, we discuss the most important consequence of Eqs. (58) and (59). It is easy to prove that the operator (59) does not commute with the electron energy operator, taken in the absence of the field (3). Therefore, from the first point of view, it seems that the equivalence between electron passive gravitational mass and its energy is broken even for macroscopic ensemble of the stationary quantum states. But, as we show below, it is not so. In particular, to demonstrate this equivalence at a macroscopic level, we consider a macroscopic ensemble of hydrogen atoms which are in a stationary quantum state with a definite energy E_n. In this case, we calculate the expectation value of the electron passive gravitational mass operator (per unit atom) from Eq. (59) in the following way:

$$\langle \hat{m}_e^g \rangle = m_e + \frac{E_n}{c^2} + \left\langle 2\frac{\hat{\mathbf{p}}^2}{2m_e} - \frac{e^2}{r} \right\rangle \bigg/ c^2 = m_e + \frac{E_n}{c^2}. \tag{60}$$

(Note that the third (virial) term in Eq. (60) is zero, according to the quantum virial theorem.[11]) As a result of the calculations, using the quantum virial theorem, we can conclude that the equivalence between passive gravitational mass and energy survives at a macroscopic level for the stationary quantum states in the non-relativistic approximation. We point out the important difference between our quantum results[7,9] of Eq. (60) and the corresponding classical ones[12] is that the expectation value of the passive gravitational mass corresponds to averaging procedure over a macroscopic ensemble of the atoms, whereas, in classical case, one averages over time.

5.2. Relativistic corrections to a hydrogen atom

In this section, we study a more general Hamiltonian, which takes into account the relativistic corrections to electron wave functions in a hydrogen atom. It is known that there are three corrections, which can be derived from the Dirac equation and which have different physical meanings.[15] As a result, the total relativistic Hamiltonian in the absence of gravitational field can be represented as

$$\hat{H}(\hat{\mathbf{p}}, \mathbf{r}) = \hat{H}_0(\hat{\mathbf{p}}, \mathbf{r}) + \hat{H}_1(\hat{\mathbf{p}}, \mathbf{r}), \tag{61}$$

with the following correction terms,

$$\hat{H}_1(\hat{\mathbf{p}}, \mathbf{r}) = \alpha \hat{\mathbf{p}}^4 + \beta \delta^3(\mathbf{r}) + \gamma \frac{\hat{\mathbf{S}} \cdot \hat{\mathbf{L}}}{r^3}, \tag{62}$$

where the parameters α, β, and γ are as follows:

$$\alpha = -\frac{1}{8m_e^3 c^2}, \quad \beta = \frac{\pi e^2 \hbar^2}{2m_e^2 c^2}, \quad \gamma = \frac{e^2}{2m_e^2 c^2}. \tag{63}$$

Here, we discuss the physical meaning of the relativistic corrections. Note that the first contribution in Eq. (62) is called the kinetic term and follows from the relativistic relation between energy and momentum. In the second correction term, which has a complicated physical meaning and is called Darwin's term, $\delta^3(\mathbf{r}) = \delta(x)\delta(y)\delta(z)$ is a three-dimensional Dirac's delta-function. And finally, the third relativistic correction term is the spin-orbital interaction, where $\hat{\mathbf{L}} = -i\hbar[\mathbf{r} \times \partial/\partial \mathbf{r}]$ is electron angular momentum operator. Then, in the presence of the weak gravitational field (3), the Schrödinger equation for electron wave functions in the local proper space-time coordinates (4), in the absence of the tidal effects, can be approximately written as

$$i\hbar \frac{\partial \Psi(\tilde{\mathbf{r}}, \tilde{t})}{\partial \tilde{t}} = [\hat{H}_0(\hat{\tilde{\mathbf{p}}}, \tilde{\mathbf{r}}) + \hat{H}_1(\hat{\tilde{\mathbf{p}}}, \tilde{\mathbf{r}})]\Psi(\tilde{\mathbf{r}}, \tilde{t}). \tag{64}$$

Using the coordinates transformation (4), the corresponding relativistic Hamiltonian in the inertial coordinate system (t, x, y, z) can be expressed as

$$\hat{H}(\hat{\mathbf{p}}, \mathbf{r}) = [\hat{H}_0(\hat{\mathbf{p}}, \mathbf{r}) + \hat{H}_1(\hat{\mathbf{p}}, \mathbf{r})]\left(1 + \frac{\phi}{c^2}\right)$$
$$+ \left(2\frac{\hat{\mathbf{p}}^2}{2m_e} - \frac{e^2}{r} + 4\alpha\hat{\mathbf{p}}^4 + 3\beta\delta^3(\mathbf{r}) + 3\gamma\frac{\hat{\mathbf{S}} \cdot \hat{\mathbf{L}}}{r^3}\right)\frac{\phi}{c^2}. \tag{65}$$

For the relativistic Hamiltonian equation (65), the operator of passive gravitational mass of electron can be written in more complicated form than that in Eq. (59):

$$\hat{m}_e^g = m_e + \left(\frac{\hat{\mathbf{p}}^2}{2m_e} - \frac{e^2}{r} + \alpha\hat{\mathbf{p}}^4 + \beta\delta^3(\mathbf{r}) + \gamma\frac{\hat{\mathbf{S}} \cdot \hat{\mathbf{L}}}{r^3}\right)\bigg/ c^2$$
$$+ \left(2\frac{\hat{\mathbf{p}}^2}{2m_e} - \frac{e^2}{r} + 4\alpha\hat{\mathbf{p}}^4 + 3\beta\delta^3(\mathbf{r}) + 3\gamma\frac{\hat{\mathbf{S}} \cdot \hat{\mathbf{L}}}{r^3}\right)\bigg/ c^2. \tag{66}$$

Below, we consider one more time a macroscopic ensemble of the hydrogen atoms, with each of them being in a stationary quantum state with a definite energy E_n'. We note that now E_n' takes into account the relativistic corrections (62) to electron energy. In this case, the expectation value of the relativistic electron mass operator (66) per atom is as follows:

$$\langle \hat{m}_e^g \rangle = m_e + \frac{E_n'}{c^2} + \left\langle 2\frac{\hat{\mathbf{p}}^2}{2m_e} - \frac{e^2}{r} + 4\alpha\hat{\mathbf{p}}^4 + 3\beta\delta^3(\mathbf{r}) + 3\gamma\frac{\hat{\mathbf{S}}\cdot\hat{\mathbf{L}}}{r^3} \right\rangle \Big/ c^2. \tag{67}$$

We stress that the Einstein's Equivalence Principle will survive at a macroscopic level for the stationary quantum states if the expectation value of the quantum virial term in Eq. (67) is zero. Below, we demonstrate, therefore, that the Einstein's equation, related the expectation value of passive gravitational mass and energy, can be applied to the stationary quantum states. To this end, we define the so-called virial operator,[11]

$$\hat{G} = \frac{1}{2}(\hat{\mathbf{p}}\mathbf{r} + \mathbf{r}\hat{\mathbf{p}}), \tag{68}$$

and make use of the equation of motion for its expectation value:

$$\frac{d}{dt}\langle \hat{G} \rangle = \frac{i}{\hbar}\left\langle [\hat{H}_0(\hat{\mathbf{p}}, \mathbf{r}) + H_1(\hat{\mathbf{p}}, \mathbf{r}), \hat{G}] \right\rangle, \tag{69}$$

where $[\hat{A}, \hat{B}]$, as usual, denotes a commutator of two operators, $\hat{A}$ and $\hat{B}$. Note that, in Eq. (69), the derivative $d\langle \hat{G} \rangle/dt$ has to be zero, since we consider the stationary quantum state. Thus,

$$\langle [\hat{H}_0(\hat{\mathbf{p}}, \mathbf{r}) + H_1(\hat{\mathbf{p}}, \mathbf{r}), \hat{G}] \rangle = 0, \tag{70}$$

where the Hamiltonian $\hat{H}_0(\hat{\mathbf{p}}, \mathbf{r}) + \hat{H}_1(\hat{\mathbf{p}}, \mathbf{r})$ is defined by Eqs. (62) and (64). At this point, using rather lengthy but straightforward calculations, we show that

$$\frac{[\hat{H}_0(\hat{\mathbf{p}}, \mathbf{r}), \hat{G}]}{-i\hbar} = 2\frac{\hat{\mathbf{p}}^2}{2m_e} - \frac{e^2}{r}, \quad \frac{[\alpha\hat{\mathbf{p}}^4, \hat{G}]}{-i\hbar} = 4\alpha\hat{\mathbf{p}}^4,$$

$$\frac{[\beta\delta^3(\mathbf{r}), \hat{G}]}{-i\hbar} = 3\beta\delta^3(\mathbf{r}), \quad \frac{1}{-i\hbar}\left[\gamma\frac{\hat{\mathbf{S}}\cdot\hat{\mathbf{L}}}{r^3}, \hat{G}\right] = 3\gamma\frac{\hat{\mathbf{S}}\cdot\hat{\mathbf{L}}}{r^3}, \tag{71}$$

where we take account of the following equality:

$$x_i\frac{d[\delta(x_i)]}{dx_i} = -\delta(x_i). \tag{72}$$

From Eqs. (70) and (71), it follows that

$$\left\langle 2\frac{\hat{\mathbf{p}}^2}{2m_e} - \frac{e^2}{r} + 4\alpha\hat{\mathbf{p}}^4 + 3\beta\delta^3(\mathbf{r}) + 3\gamma\frac{\hat{\mathbf{S}}\cdot\hat{\mathbf{L}}}{r^3}\right\rangle = 0, \qquad (73)$$

and, thus, Eq. (67) can be represented in the standard Einstein's form:

$$\langle \hat{m}_e^g \rangle = m_e + \frac{E'_n}{c^2}. \qquad (74)$$

Let us discuss the status of the Einstein's Equivalence Principle for the stationary quantum states, considered in this section. Note that Eq. (74) directly establishes the equivalence between the expectation value of electron passive gravitational mass and its energy in a hydrogen atom, including the relativistic corrections. Thus, we can say that the Equivalence Principle survives for quantum macroscopic bodies, which contains quantum composite bodies in the stationary quantum states, which first shown by us in Refs. 7 and 9. On the other hand, for stationary quantum states the Einstein's Equivalence Principle is broken at microscopic level. Indeed, as first shown by us in Refs. 7 and 9 (see also Secs. 3 and 4), the quantum measurement of passive gravitational mass in state with a definite energy, E_n, can give with small probabilities the values $m_e^g \neq m_e + \frac{E_n}{c^2}$. Although, we consider in this chapter the simplest quantum composite body — a hydrogen atom, we speculate that our results survive also for more complicated quantum systems, including many-body systems with arbitrary interactions of particles.

6. Inequivalence Between Passive Gravitational Mass and Energy at a Macroscopic Level for the Gravitational Demons (Method-1)

Below, we use the accepted in this chapter procedure of the quantum measurements of passive gravitational mass for a macroscopic ensemble. It is important that the expectation values of energy and gravitational mass have to be calculated at the same moment of time, $t = t' = 0$. We suggest that, at $t < 0$, where the gravitational field is absent, we have a macroscopic ensemble of coherent superpositions of two wave functions, corresponding to the ground-state ($1S$) wave function, $\Psi_1(r)$, and the first excited energy-level ($2S$) wave function, $\Psi_2(r)$, in a hydrogen atom:

$$\Psi(r, t) = \frac{1}{\sqrt{2}}\exp\left(\frac{-im_e c^2 t}{\hbar}\right)\left[\exp\left(\frac{-iE_1 t}{\hbar}\right)\Psi_1(r) + \exp\left(\frac{-iE_2 t}{\hbar}\right)\Psi_2(r)\right],$$
$$(75)$$

where we omit the $2P$ wave function due to the fact that the wave functions with different parities do not mix in the gravitational field (3).[7,9] (It is important that a macroscopic coherent ensemble of such wave functions, where the difference between phases of functions $\Psi_1(r)$ and $\Psi_2(r)$ is fixed, is possible to create by means of some laser technique.[17]) Note that the expectation value of energy in a macroscopic ensemble (75) in the absence of gravitational field is constant and is equal to

$$\langle E(t < 0) \rangle = \frac{(E_1 + E_2)}{2}. \tag{76}$$

We perform the following Gedanken experiment: we suddenly switch on the gravitational field (3) at $t = 0$. In this case, we can write the following time-dependent perturbation in the field to electron Hamiltonian:

$$U_1(\mathbf{r}, t) = \frac{\phi}{c^2}[m_e c^2 + \hat{H}_0(\mathbf{r}) + \hat{V}(\mathbf{r})]\Theta(t), \tag{77}$$

where $\Theta(t)$ is the step-function and, as shown in Eq. (59), the virial term can be represent as

$$\hat{V}(\mathbf{r}) = 2\frac{\hat{\mathbf{p}}^2}{2m} - \frac{e^2}{r}. \tag{78}$$

As we already mentioned, we disregard all small probabilities of the order of $\frac{\phi^2}{c^4}$ for electron to be in the gravitational field (3) in energy level with $n > 2$. Thus, in a hydrogen atom we can consider only two levels, $n = 1$ and $n = 2$, and apply to them the two-level variant of the time-dependent perturbation theory.[11] In accordance with this variant, in the gravitational field (3), the wave function can be represented as follows:

$$\Psi^1(r, t) = \exp\left(\frac{-im_e c^2 t}{\hbar}\right)\left[\exp\left(\frac{-iE_1 t}{\hbar}\right)a_1(t)\Psi_1(r)\right.$$

$$\left. + \exp\left(\frac{-iE_2 t}{\hbar}\right)a_2(t)\Psi_2(r)\right]. \tag{79}$$

By means of the standard quantum time-dependent perturbation theory, we can obtain the following equations to determine the functions $a_1(t)$ and

$a_2(t)$ in Eq. (79):

$$\frac{da_1(t)}{dt} = -i\,U_{11}(t)\,a_1(t) - i\,U_{12}(t)\exp\left[-i\frac{(E_2 - E_1)t}{\hbar}\right]a_2(t),$$

$$\frac{da_2(t)}{dt} = -i\,U_{22}(t)\,a_2(t) - i\,U_{21}(t)\exp\left[-i\frac{(E_1 - E_2)t}{\hbar}\right]a_1(t), \quad (80)$$

with the matrix elements of the perturbation (77),(78) being:

$$U_{11}(t) = \Theta(t)\frac{\phi}{c^2}\int \Psi_1^*(r)[m_ec^2 + \hat{H}_0(\mathbf{r}) + \hat{V}(\mathbf{r})]\Psi_1(r)d^3\mathbf{r}$$

$$= \Theta(t)\frac{\phi}{c^2}(m_ec^2 + E_1),$$

$$U_{12}(t) = \Theta(t)\frac{\phi}{c^2}\int \Psi_1^*(r)[m_ec^2 + \hat{H}_0(\mathbf{r}) + \hat{V}(\mathbf{r})]\Psi_2(r)d^3\mathbf{r}$$

$$= \Theta(t)\frac{\phi}{c^2}V_{12}, \quad (81)$$

$$U_{22}(t) = \Theta(t)\frac{\phi}{c^2}\int \Psi_2^*(r)[m_ec^2 + \hat{H}_0(\mathbf{r}) + \hat{V}(\mathbf{r})]\Psi_2(r)d^3\mathbf{r}$$

$$= \Theta(t)\frac{\phi}{c^2}(m_ec^2 + E_2),$$

$$U_{21}(t) = \Theta(t)\frac{\phi}{c^2}\int \Psi_2^*(r)[m_ec^2 + \hat{H}_0(\mathbf{r}) + \hat{V}(\mathbf{r})]\Psi_1(r)d^3\mathbf{r}$$

$$= \Theta(t)\frac{\phi}{c^2}V_{21},$$

where V_{ij} are the matrix elements of the virial operator (78). After direct solving of Eqs.(80) and (81), it is possible to find that the function (79) is

$$\Psi^1(r,t) = \exp\left(\frac{-im_ec^2t}{\hbar}\right)\left[\Psi_1^1(r,t) + \Psi_2^1(r,t)\right], \quad (82)$$

where

$$\Psi_1^1(r,t) = \frac{1}{\sqrt{2}} \exp\left[-i\frac{(m_ec^2 + E_1)\phi\, t}{c^2\hbar}\right]$$

$$\times \exp\left(-i\frac{E_1 t}{\hbar}\right)\left[1 - \frac{\phi V_{12}}{c^2(E_2 - E_1)}\right]\Psi_1(r)$$

$$+ \frac{1}{\sqrt{2}}\exp\left(-i\frac{E_2 t}{\hbar}\right)\frac{\phi V_{12}}{c^2(E_2 - E_1)}\Psi_1(r) \tag{83}$$

and

$$\Psi_2^1(r,t) = \frac{1}{\sqrt{2}} \exp\left[-i\frac{(m_ec^2 + E_2)\phi\, t}{c^2\hbar}\right]$$

$$\times \exp\left(-i\frac{E_2 t}{\hbar}\right)\left[1 - \frac{\phi V_{21}}{c^2(E_1 - E_2)}\right]\Psi_2(r)$$

$$+ \frac{1}{\sqrt{2}}\exp\left(-i\frac{E_1 t}{\hbar}\right)\frac{\phi V_{21}}{c^2(E_1 - E_2)}\Psi_2(r). \tag{84}$$

It is easy to show that with the excepted accuracy the wave function (82)–(84) can be written in the following more convenient way:

$$\Psi^1(r,t) = \frac{1}{\sqrt{2}}\exp\left[-i\frac{(m_ec^2 + E_1)(1+\phi/c^2)t}{\hbar}\right]\left\{\left[1 - \frac{\phi V_{12}}{c^2(E_2 - E_1)}\right]\Psi_1(r)\right.$$

$$+ \frac{\phi V_{21}}{c^2(E_1 - E_2)}\Psi_2(r)\right\} + \frac{1}{\sqrt{2}}\exp\left[-i\frac{(m_ec^2 + E_2)(1+\phi/c^2)t}{\hbar}\right]$$

$$\times\left\{\left[1 - \frac{\phi V_{21}}{c^2(E_1 - E_2)}\right]\Psi_2(r) + \frac{\phi V_{12}}{c^2(E_2 - E_1)}\Psi_1(r)\right\}, \tag{85}$$

where the wave function (85), taken with the same accuracy, is normalized:

$$\int [\Psi^1(r,t)]^*\Psi^1(r,t)d^3r \approx 1. \tag{86}$$

As we wrote, wave functions in a macroscopic ensemble of the quantum coherent superpositions at $t = t' = 0$ (see Eq. (75)) are characterized by the constant phase shift and, therefore, can be written as

$$\Psi(r,t) = \frac{1}{\sqrt{2}} \exp\left(\frac{-im_e c^2 t}{\hbar}\right)\left[\exp\left(\frac{-iE_1 t}{\hbar}\right)\Psi_1^0(r)\right.$$

$$\left. + \exp(i\tilde{\alpha})\exp\left(\frac{-iE_2 t}{\hbar}\right)\Psi_2^0(r)\right], \quad (87)$$

with $\Psi_1^0(r)$ and $\Psi_2^0(r)$ being the corresponding real functions and relative phase $\tilde{\alpha}$ being constant. In this particular case, after some calculations, it is possible to show that the expectation value of energy in the quantum state (75) in the weak gravitational field (3) is

$$\langle E(t \geq 0)\rangle = \int [\Psi^1(r,t)]^* \left(i\hbar \frac{\partial}{\partial t}\right)\Psi^1(r,t)d^3 r$$

$$= m_e c^2\left(1 + \frac{\phi}{c^2}\right) + \frac{(E_1 + E_2)}{2}\left(1 + \frac{\phi}{c^2}\right) + \tilde{V}_{12}\frac{\phi}{c^2}\cos(\tilde{\alpha}), \quad (88)$$

where

$$\tilde{V}_{12} = \int \Psi_1^0(r)\hat{V}(\mathbf{r})\Psi_2^0(r)d^3 r = \int \Psi_1^0(r)\left(2\frac{\hat{p}^2}{2m} - \frac{e^2}{r}\right)\Psi_2^0(r)d^3 r. \quad (89)$$

From Eq. (88), it is clear that macroscopic ensemble of the coherent superpositions of quantum states (75),(87) (i.e. Gravitational demon) is characterized by the following expectation value of electron gravitational mass per one hydrogen atom:

$$\langle m_e^g \rangle = m_e + \frac{(E_1 + E_2)}{2c^2} + \frac{\tilde{V}_{12}}{c^2}\cos(\tilde{\alpha}). \quad (90)$$

Note that here m_e is the bare electron mass; the second term is the expected kinetic and potential energy contributions to gravitational mass, whereas the third virial term is non-trivial virial contribution to electron mass. Therefore, Eq. (90) explicitly demonstrates inequivalence between the expectation values of energy (76) and gravitational mass of a macroscopic ensemble of the coherent superpositions of the stationary states. Note that, if we have the incoherent ensemble (where the phase $\tilde{\alpha}$ is not fixed), then the gravitational mass (90) quickly oscillates with oscillating phase and the equivalence between the expectation values (76) and (90) restores. For the coherent macroscopic ensemble, Eq. (90) crucially depends on the ensemble preparation procedure (i.e. on the phase shift $\tilde{\alpha}$). For instance, the expectation value of the gravitational mass can be both larger and smaller than the expected value from the Einstein's Equivalence Principle and for the simple cases $\tilde{\alpha}_1 = 0$, $\tilde{\alpha}_2 = \pi/2$, and $\tilde{\alpha}_3 = \pi$ is equal:

$$\langle m_e^g \rangle = m_e + \frac{(E_1 + E_2)}{2c^2} + \frac{\tilde{V}_{12}}{c^2}, \quad \tilde{\alpha}_1 = 0, \tag{91}$$

$$\langle m_e^g \rangle = m_e + \frac{(E_1 + E_2)}{2c^2}, \quad \tilde{\alpha}_2 = \pi/2, \tag{92}$$

and

$$\langle m_e^g \rangle = m_e + \frac{(E_1 + E_2)}{2c^2} - \frac{\tilde{V}_{12}}{c^2}, \quad \tilde{\alpha}_3 = \pi. \tag{93}$$

To make sure that the suggested effect is not zero, we have calculated the virial matrix element for a hydrogen atom and have found that

$$\tilde{V}_{12} = 0.56 \, (E_2 - E_1). \tag{94}$$

7. Inequivalence Between Passive Gravitational Mass and Energy at a Macroscopic Level for the Gravitational Demons (Method-2)

In this section, we make use of different method to obtain electron passive gravitational mass (90). As shown in Refs. 13 and 14, in classical case the virial term disappears if we introduce the proper local coordinates $(\tilde{x}, \tilde{y}, \tilde{z})$ (see Eq. (4)). Therefore, it is important to show that Eq. (90) survives in quantum case if we consider the problem in the proper local coordinates (4). Here, to measure gravitational mass, we perform the same Gedanken experiment, as in Sec. 6. In particular, at $t < 0$, there is no gravitational field and we have a macroscopic ensemble of the coherent superpositions of two stationary wave functions (see Eq. (75)). We switch on gravitational field (3) at $t = 0$, which is equivalent to a change of geometry of the space-time. According to the property of the local Lorentz invariance, which is a part of the Einstein's Equivalence Principle, the solution of the Schrödinger equation at $t \geq 0$ can be written in the local proper coordinates (4) as

$$\Psi_2(\tilde{r}, \tilde{t}) = \exp\left(\frac{-im_e c^2 \tilde{t}}{\hbar}\right) \left[A \exp\left(\frac{-iE_1 \tilde{t}}{\hbar}\right) \Psi_1^0(\tilde{r}) \right.$$

$$\left. + B \exp\left(\frac{-iE_2 \tilde{t}}{\hbar}\right) \Psi_2^0(\tilde{r}) \right], \tag{95}$$

where complex parameters A and B are not necessarily equal to $\frac{1}{\sqrt{2}}$ and take into account the gravitational field (3). On the other hand, in the proper local coordinates, we can represent the wave function (75) as

$$
\Psi_3(\tilde{r}, \tilde{t}) = \frac{1}{\sqrt{2}} \exp\left[\frac{-im_e c^2 \tilde{t}(1 - \phi/c^2)}{\hbar}\right] \left(1 + \frac{\phi}{c^2}\right)^{3/2}
$$

$$
\times \left\{ \exp\left[\frac{-iE_1 \tilde{t}(1 - \phi/c^2)}{\hbar}\right] \Psi_1^0[\tilde{r}(1 + \phi/c^2)] \right.
$$

$$
\left. + \exp(i\tilde{\alpha}) \exp\left[\frac{-iE_2 \tilde{t}(1 - \phi/c^2)}{\hbar}\right] \Psi_2^0[\tilde{r}(1 + \phi/c^2)] \right\}. \quad (96)
$$

(Note that the wave function (95) is normalized in the proper local coordinates (4).) It is obvious that, at $t = t' = 0$, the above-discussed wave functions (95) and (96) have to be equal to each other:

$$
\Psi_2(\tilde{r}, \tilde{t} = 0) = \Psi_3(\tilde{r}, \tilde{t} = 0). \quad (97)
$$

Using (97) and the following orthogonality condition for real functions $\Psi_1^0(\tilde{r})$ and $\Psi_2^0(\tilde{r})$,

$$
\int_0^\infty [\Psi_1^0(\tilde{r})]^2 d^3\tilde{r} = \int_0^\infty [\Psi_2^0(\tilde{r})]^2 d^3\tilde{r} = 1, \quad \int_0^\infty \Psi_1^0(\tilde{r})\Psi_2^0(\tilde{r}) d^3\tilde{r} = 0,
$$

$$
(98)
$$

it is possible to define the parameters A and B in Eq. (95):

$$
A = \frac{1}{\sqrt{2}}[\Delta_{11} + \exp(i\tilde{\alpha})\Delta_{12}], \quad B = \frac{1}{\sqrt{2}}[\Delta_{21} + \exp(i\tilde{\alpha})\Delta_{22}], \quad (99)
$$

where we introduce the following matrix:

$$
\hat{\Delta}_{ij} = \left(1 + \frac{\phi}{c^2}\right)^{3/2} \int_0^\infty \Psi_i^0(\tilde{r})\Psi_j^0\left[\tilde{r}\left(1 + \frac{\phi}{c^2}\right)\right] d^3\tilde{r}. \quad (100)
$$

Let us calculate the matrix elements of the $\hat{\Delta}_{ij}$ with the accepted, in this chapter, accuracy — to the first order of the small parameter $|\frac{\phi}{c^2}| \ll 1$.

By the definition (100),

$$\Delta_{11} = \left(1 + \frac{\phi}{c^2}\right)^{3/2} \int_0^\infty \Psi_1^0(\tilde{r})\Psi_1^0\left[\tilde{r}\left(1 + \frac{\phi}{c^2}\right)\right]d^3\tilde{r}$$

$$\approx \left(1 - \frac{\phi}{c^2}\right)^{3/2} \int_0^\infty \Psi_1^0\left[r\left(1 - \frac{\phi}{c^2}\right)\right]\Psi_1^0(r)d^3r. \qquad (101)$$

From Eq. (101), it follows that Δ_{11} is an even function of the variable $\frac{\phi}{c^2}$, therefore, in our approximation

$$\Delta_{11} = \int_0^\infty [\Psi_1^0(\tilde{r})]^2 d^3\tilde{r} = 1. \qquad (102)$$

Using the same method, it is easy to show that

$$\Delta_{22} = \int_0^\infty [\Psi_2^0(\tilde{r})]^2 d^3\tilde{r} = 1. \qquad (103)$$

Calculations of non-diagonal matrix elements of the matrix $\hat{\Delta}_{ij}$ are more complicated procedures. Let us start from calculation of the matrix element Δ_{12}. From (100), we have

$$\Delta_{12} = \left(1 + \frac{\phi}{c^2}\right)^{3/2} \int_0^\infty \Psi_1^0(\tilde{r})\Psi_2^0\left[\tilde{r}\left(1 + \frac{\phi}{c^2}\right)\right]d^3\tilde{r}$$

$$\approx \left(1 - \frac{\phi}{c^2}\right)^{3/2} \int_0^\infty \Psi_1^0\left[r\left(1 - \frac{\phi}{c^2}\right)\right]\Psi_2^0(r)d^3r$$

$$\approx \left(1 - \frac{\phi}{c^2}\right)^{3/2} \int_0^\infty \Psi_1^0(r)\Psi_2^0(r)d^3r$$

$$- \frac{\phi}{c^2}\int_0^\infty \Psi_1'(r)\ r\ \Psi_2^0(r)d^3r = -\frac{\phi}{c^2}\ \Delta, \qquad (104)$$

where

$$\Psi_1'(r) = \frac{d\Psi_1^0(r)}{dr}, \quad \Delta = \int_0^\infty \Psi_1'(r)\ r\ \Psi_2^0(r)d^3r. \qquad (105)$$

Using the same method, we can find that

$$\Delta_{21} = +\frac{\phi}{c^2}\ \Delta. \qquad (106)$$

For further development, it is necessary to calculate quantity Δ in Eqs. (104)–(106) in terms of the virial contribution. In particular, let us show that

$$\Delta = \int_0^\infty \Psi_1'(r)\, r\, \Psi_2^0(r) d^3 r = \frac{\tilde{V}_{12}}{E_2 - E_1}, \tag{107}$$

where the matrix elements of the virial operator (78), V_{12}, are defined by the equation:

$$\tilde{V}_{12} = \int_0^\infty \Psi_1^0(r)\hat{V}(r)\Psi_2^0(r) d^3 r. \tag{108}$$

To this end, we rewrite the Schrödinger equation in gravitational field (3) in terms of the unperturbed space-time coordinates (t, x, y, z) (4):

$$(m_e c^2 + E_1)\Psi_1^0\left[\left(1 - \frac{\phi}{c^2}\right)r\right] = \left[m_e c^2 - \frac{1}{(1 - \phi/c^2)^2}\frac{\hbar^2}{2m_e}\right.$$
$$\times \left(\frac{\partial^2}{\partial x^2} + \frac{\partial^2}{\partial y^2} + \frac{\partial^2}{\partial z^2}\right)$$
$$\left. - \frac{1}{(1 - \phi/c^2)}\frac{e^2}{r}\right]\Psi_1^0\left[\left(1 - \frac{\phi}{c^2}\right)r\right]. \tag{109}$$

Then, using the weak field approximation and, thus, keeping only terms of the first order with respect to the small parameter $|\frac{\phi}{c^2}| \ll 1$, we obtain

$$E_1\Psi_1^0(r) - \frac{\phi}{c^2} E_1\, r\, \Psi_1'(r) = \left[-\frac{\hbar^2}{2m_e}\left(\frac{\partial^2}{\partial x^2} + \frac{\partial^2}{\partial y^2} + \frac{\partial^2}{\partial z^2}\right) - \frac{e^2}{r} + \frac{\phi}{c^2}\,\hat{V}(r)\right]$$
$$\times \left[\Psi_1^0(r) - \frac{\phi}{c^2}\, r\, \Psi_1'(r)\right]. \tag{110}$$

As follows from (110),

$$-E_1 r\Psi_1'(r) = \left[\frac{\hbar^2}{2m_e}\left(\frac{\partial^2}{\partial x^2} + \frac{\partial^2}{\partial y^2} + \frac{\partial^2}{\partial z^2}\right) + \frac{e^2}{r}\right]r\Psi_1'(r) + \hat{V}(r)\Psi_1^0(r). \tag{111}$$

Let us multiply Eq. (111) on $\Psi_2^0(r)$ and integrate over space,

$$-E_1\int_0^\infty \Psi_2^0(r)r\Psi_1'(r)d^3 r = \int_0^\infty \Psi_2^0(r)\left[\frac{\hbar^2}{2m_e}\left(\frac{\partial^2}{\partial x^2} + \frac{\partial^2}{\partial y^2} + \frac{\partial^2}{\partial z^2}\right) + \frac{e^2}{r}\right]$$
$$\times r\Psi_1'(r)d^3 r + \int_0^\infty \Psi_2^0(r)\hat{V}(r)\Psi_1^0(r)d^3 r. \tag{112}$$

If we make use of the fact that the Hamiltonian and virial term are the Hermitian operators, we can rewrite (112) as

$$E_1 \int_0^\infty \Psi_2^0(r) r \Psi_1'(r) d^3r = E_2 \int_0^\infty \Psi_2^0(r) r \Psi_1'(r) d^3r$$

$$- \int_0^\infty \Psi_1^0(r) \hat{V}(r) \Psi_2^0(r) d^3r. \qquad (113)$$

Then, Eq. (107) directly follows from (113):

$$\Delta = \int_0^\infty \Psi_1'(r) r \Psi_2(r) d^3r = \frac{\tilde{V}_{1,2}}{E_2 - E_1}. \qquad (114)$$

First, let us check that the wave function (95) is normalized in the proper local coordinates (4) with the necessary accuracy. To this end, we calculate

$$\int_0^\infty |\Psi_2(0, \tilde{r})|^2 d^3\tilde{r} = \int_0^\infty [A^* \Psi_1^0(\tilde{r}) + B^* \Psi_2^0(\tilde{r})][A\Psi_1^0(\tilde{r}) + B\Psi_2^0(\tilde{r})] d^3\tilde{r}$$

$$= |A|^2 + |B|^2, \qquad (115)$$

where

$$|A|^2 + |B|^2 = \frac{1}{2}\left\{\left[1 - \frac{\phi}{c^2}\Delta \exp(-i\tilde{\alpha})\right]\left[1 - \frac{\phi}{c^2}\Delta \exp(+i\tilde{\alpha})\right]\right.$$

$$\left. + \left[\exp(-i\tilde{\alpha}) + \frac{\phi}{c^2}\Delta\right]\left[\exp(+i\tilde{\alpha}) + \frac{\phi}{c^2}\Delta\right]\right\} \approx 1 \quad (116)$$

(see Eqs. (99), (104) and (106)).

Second, let us calculate the expectation value of energy for wave function (95) in the gravitational field (3):

$$\langle E \rangle = |A|^2 (E_1 + m_e c^2)\left(1 + \frac{\phi}{c^2}\right) + |B|^2 (E_2 + m_e c^2)\left(1 + \frac{\phi}{c^2}\right)$$

$$= \frac{1}{2}\left\{(E_1 + m_e c^2)\left(1 + \frac{\phi}{c^2}\right)\left[1 - \frac{\phi}{c^2}\Delta \exp(-i\tilde{\alpha})\right]\left[1 - \frac{\phi}{c^2}\Delta \exp(+i\tilde{\alpha})\right]\right.$$

$$\left. + (E_2 + m_e c^2)\left(1 + \frac{\phi}{c^2}\right)\left[\exp(-i\tilde{\alpha}) + \frac{\phi}{c^2}\Delta\right]\left[\exp(+i\tilde{\alpha}) + \frac{\phi}{c^2}\Delta\right]\right\}$$

$$\approx m_e c^2 + \frac{1}{2}(E_1 + E_2) + m_e\phi + \frac{1}{2c^2}(E_1 + E_2)\phi + \frac{\tilde{V}_{12}}{c^2}\cos(\tilde{\alpha})\phi. \quad (117)$$

The equation for the expectation value of gravitational electron mass (90) directly follows from Eq. (117) and contains the discussed above virial contribution. So in this section, we have derived Eq. (90), using the local proper coordinates (4), in contrast to the classical case statements.[13, 14]

7.1. *Some experimental aspects*

From our results, discussed in the chapter, it follows that in order to discover breakdown of the Einstein's Equivalence Principle it is not necessarily to perform long and expensive experiments in the space. What we actually need is to create special macroscopic ensemble of the coherent superpositions of two or several stationary quantum states (which we call Gravitational demon) and to measure its weight. It is possible to create such macroscopic ensemble, using some laser technology (see Ref. 17). Then, it is necessary to measure the Gravitation demon's energy and to compare these two quantities. To evaluate energy in the considered case, it seems to be a good idea just to count a number of the emitted photons from the macroscopic ensemble of the atoms. Let us discuss some obvious difficulties. It is evident that weight of the created Gravitational demon has to be measured at the moment of time which is very close to its creation. In this chapter, we have used for introduction of the gravitational field the so-called step-like function, $\Theta(t)$. Of course, this does not mean a motion in the gravitational field (3) with speed higher than the speed of light. We can use step-like function, if significant change of the gravitational field happens quicker than the characteristic period of quasi-classical rotation of electrons in a hydrogen atom. More strictly speaking, in our case of superposition of two atomic levels, we need the time of the order of $\delta t \sim t_0 = \frac{2\pi \hbar}{E_2 - E_1} \sim 10^{-15}$ s. To conclude, we have demonstrated, using Gedanken experiments at the Earth's laboratory, that breakdown of the Einstein's Equivalence Principle is possible for some special macroscopic quantum states. Finding the best concrete realizations for the corresponding real experiments is the topic of our future investigations.

8. Inequivalence Between Active Gravitational Mass and Energy at a Macroscopic Level for the Gravitational Demons

Below, we discuss our results,[8, 9] where we showed that active gravitational mass and energy were inequivalent to each other at macroscopic level for coherent ensembles of quantum superpositions of stationary states (i.e. for Gravitational demons).

8.1. *Active gravitational mass in classical physics*

Here, we determine electron active gravitational mass in a classical model of a hydrogen atom, which takes into account electron kinetic and potential Coulomb energies.[12] More specifically, we consider a particle with small bare mass m_e, moving in the Coulomb electrostatic field of a heavy particle with bare mass $m_p \gg m_e$. Our task is to find gravitational potential at large distance from the atom, $R \gg r_B$, where r_B is the the so-called Bohr radius (i.e. effective "size" of a hydrogen atom). Here, we use the so-called weak field gravitational theory,[1, 12] where the post-Newtonian gravitational potential can be written as[8, 9]

$$\phi(R,t) = -G\frac{m_p + m_e}{R} - G \int \frac{\Delta T_{\alpha\alpha}^{kin}(t,\mathbf{r}) + \Delta T_{\alpha\alpha}^{pot}(t,\mathbf{r})}{c^2 R} d^3\mathbf{r}, \qquad (118)$$

where $\Delta T_{\alpha\beta}^{kin}(t,\mathbf{r})$ and $\Delta T_{\alpha\beta}^{pot}(t,\mathbf{r})$ are contributions to stress–energy tensor density, $T_{\alpha\beta}(t,\mathbf{r})$, due to kinetic and the Coulomb potential energies, respectively. We stress that, in Eq. (118), we disregard all retardation effects. Thus, in the above-discussed approximation, electron active gravitational mass is equal to

$$m_e^a = m_e + \frac{1}{c^2} \int [\Delta T_{\alpha\alpha}^{kin}(t,\mathbf{r}) + \Delta T_{\alpha\alpha}^{pot}(t,\mathbf{r})] d^3\mathbf{r}. \qquad (119)$$

Here, we calculate $\Delta T_{\alpha\alpha}^{kin}(t,\mathbf{r})$, using the standard expression for stress–energy tensor density of a moving relativistic point mass[1, 2]:

$$T_{kin}^{\alpha\beta}(\mathbf{r},t) = \frac{m_e v^\alpha(t) v^\beta(t)}{\sqrt{1 - v^2(t)/c^2}} \, \delta^3[\mathbf{r} - \mathbf{r}_e(t)], \qquad (120)$$

where v^α is a four-velocity, $\delta^3(\ldots)$ is the three-dimensional Dirac δ-function, and $\mathbf{r}_e(t)$ is a three-dimensional electron trajectory. From Eqs. (119) and (120), it follows that

$$\Delta T_{\alpha\alpha}^{kin}(t) = \int \Delta T_{\alpha\alpha}^{kin}(t,\mathbf{r}) d^3\mathbf{r} = \frac{m_e[c^2 + v^2(t)]}{\sqrt{1 - v^2(t)/c^2}} - m_e c^2. \qquad (121)$$

We note that, although calculations of the contribution from Coulomb potential energy to stress–energy tensor are more complicated, they can be done by using the formula for stress–energy tensor of electromagnetic field,[1]

$$T^{\mu\nu}_{em} = \frac{1}{4\pi}[F^{\mu\alpha}F^{\nu}{}_{\alpha} - \frac{1}{4}\eta^{\mu\nu}F_{\alpha\beta}F^{\alpha\beta}], \tag{122}$$

where $\eta_{\alpha\beta}$ is the Minkowski's metric tensor, $F^{\alpha\beta}$ is the so-called Maxwell tensor of electromagnetic field.[1] Here, as was mentioned above, we use approximation, where we do not take into account magnetic field and keep only the Coulomb electrostatic field. Then, we can simplify Eq. (122) and obtain from it the following expression:

$$\Delta T^{pot}_{\alpha\alpha}(t) = \int \Delta T^{pot}_{\alpha\alpha}(t,\mathbf{r})d^3\mathbf{r} = -2\frac{e^2}{r(t)}. \tag{123}$$

As directly follows from (121) and (123), electron active gravitational mass can be written in the following way:

$$m^a_e = \left[\frac{m_e c^2}{(1-v^2/c^2)^{1/2}} - \frac{e^2}{r}\right]\Big/c^2 + \left[\frac{m_e v^2}{(1-v^2/c^2)^{1/2}} - \frac{e^2}{r}\right]\Big/c^2. \tag{124}$$

Note that the first term in Eq. (124) is the expected one. Indeed, it is the total energy contribution to the mass, whereas the second term is the so-called relativistic virial one.[11] It is important that it is time-dependent. Therefore, in classical physics, active gravitational mass of a composite body depends on time too. Nevertheless, in this situation, it is possible to introduce averaged over time electron active gravitational mass. This averaging procedure results in the expected equivalence between averaged over time active gravitational mass and energy[12]:

$$\langle m^a_e \rangle_t = \left\langle \frac{m_e c^2}{(1-v^2/c^2)^{1/2}} - \frac{e^2}{r} \right\rangle_t \Big/ c^2$$

$$+ \left\langle \frac{m_e v^2}{(1-v^2/c^2)^{1/2}} - \frac{e^2}{r} \right\rangle_t \Big/ c^2 = m_e + E/c^2. \tag{125}$$

We stress that, in Eq. (125), the averaged over time virial term is zero due to the classical virial theorem. It is possible to show that for non-relativistic case our Eqs. (124) and (125) can be simplified to

$$m^a_e = m_e + \left(\frac{m_e v^2}{2} - \frac{e^2}{r}\right)\Big/c^2 + \left(2\frac{m_e v^2}{2} - \frac{e^2}{r}\right)\Big/c^2 \tag{126}$$

and

$$\langle m_e^a \rangle_t = m_e + \left\langle \frac{m_e v^2}{2} - \frac{e^2}{r} \right\rangle_t \Big/ c^2$$

$$+ \left\langle 2\frac{m_e v^2}{2} - \frac{e^2}{r} \right\rangle_t \Big/ c^2 = m_e + E/c^2. \tag{127}$$

8.2. Active gravitational mass in quantum physics

Below, we consider the so-called semiclassical gravitational theory,[18] where, in the Einstein's field equation, gravitational field is not quantized but the matter is quantized:

$$R_{\mu\nu} - \frac{1}{2}Rg_{\mu\nu} = \frac{8\pi G}{c^4}\langle \hat{T}_{\mu\nu} \rangle. \tag{128}$$

In this equation, $\langle \hat{T}_{\mu\nu} \rangle$ is the expectation value of quantum operator, corresponding to the stress–energy tensor. To make use of Eq. (128), we have to rewrite Eq. (126) for electron active gravitational mass using momentum, instead of velocity. After that, we can quantize the obtained result:

$$\hat{m}_e^a = m_e + \left(\frac{\hat{\mathbf{p}}^2}{2m_e} - \frac{e^2}{r} \right) \Big/ c^2 + \left(2\frac{\hat{\mathbf{p}}^2}{2m_e} - \frac{e^2}{r} \right) \Big/ c^2. \tag{129}$$

We stress that Eq. (129) represents electron active gravitational mass operator. As follows from (129), the expectation value of active electron gravitational mass can be expressed as

$$\langle \hat{m}_e^a \rangle = m_e + \left\langle \frac{\hat{\mathbf{p}}^2}{2m_e} - \frac{e^2}{r} \right\rangle \Big/ c^2 + \left\langle 2\frac{\hat{\mathbf{p}}^2}{2m_e} - \frac{e^2}{r} \right\rangle \Big/ c^2, \tag{130}$$

where third term is the virial one.

8.2.1. Equivalence of the expectation values of active gravitational mass and energy for stationary quantum states

In this section, we consider a macroscopic ensemble of the hydrogen atoms with each of them being in the nth energy level. For such ensemble, the expectation value of the active gravitational mass operator (130) is

$$\langle \hat{m}_e^a \rangle = m_e + \frac{E_n}{c^2}. \tag{131}$$

In Eqs. (130) and (131), we take account of that the expectation value of the quantum virial operator is equal to zero in stationary quantum states due to the quantum virial theorem.[11] Therefore, we can make the following important conclusion: in stationary quantum states, active gravitational mass of a composite quantum body is equivalent to its energy at a macroscopic level.[8,9]

8.2.2. *Inequivalence between active gravitational mass and energy for Gravitational demons*

Here, we introduce the simplest macroscopic ensemble of the coherent superpositions of the following stationary quantum states in a hydrogen atom,

$$
\Psi(r,t) = \frac{1}{\sqrt{2}} \exp\left(-i\frac{m_e c^2 t}{\hbar}\right) \left[\Psi_1^0(r) \exp\left(-i\frac{E_1 t}{\hbar}\right)\right.
$$
$$
\left. + \exp(i\alpha)\Psi_2^0(r) \exp\left(-i\frac{E_2 t}{\hbar}\right)\right], \tag{132}
$$

where $\Psi_1^0(r)$ and $\Psi_2^0(r)$ are the normalized wave functions of the ground state $(1S)$ and first excited state $(2S)$, respectively. We point out that it is possible to create the coherent superposition, where $\alpha =$const. in Eq. (132) for all macroscopic ensemble, by using lasers. As we have already discussed, we call such coherent ensembles Gravitational demons. It is easy to show that the superposition (132) corresponds to the following constant expectation value of energy in the absence of the gravitational field,

$$
\langle E \rangle = m_e c^2 + \frac{E_1 + E_2}{2}. \tag{133}
$$

On the other hand, as seen from Eq. (130), the expectation value of electron active gravitational mass operator for the wave function (132) is not constant and exhibits time-dependent oscillations:

$$
\langle \hat{m}_e^a \rangle = m_e + \frac{E_1 + E_2}{2c^2} + \frac{V_{12}}{c^2} \cos\left[\alpha + \frac{(E_1 - E_2)t}{\hbar}\right], \tag{134}
$$

where V_{12} is matrix element of the quantum virial operator,

$$
V_{12} = \int \Psi_1^0(r) \left(2\frac{\hat{\mathbf{P}}^2}{2m_e} - \frac{e^2}{r}\right) \Psi_2^0(r) \, d^3\mathbf{r} \,, \tag{135}
$$

between the considered above two stationary quantum states. It is important that the oscillations (134) and (135) directly demonstrate breakdown of the equivalence between the expectation values of active gravitational mass and energy for coherent quantum superpositions of stationary states.[8,9]

We stress that such quantum time-dependent oscillations are very general and are not restricted by the case of a hydrogen atom. They have a pure quantum origin and do not have classical analogs.

8.3. *Some experimental aspects*

Below, we suggest an idealized experiment, which, in principle, allows to observe quantum time-dependent oscillations of the expectation values of active gravitational mass (134). As mentioned before, it is possible to create a macroscopic ensemble of the coherent superpositions of electron quantum stationary states in some gas with high density. It is important that these superpositions have to be characterized by the feature that each atom (or molecule) has the same phase difference between two (or several) wave function stationary components. In this case, the macroscopic ensemble of the atoms (or molecules) generates gravitational field, which oscillates in time similar to Eq. (134), which can be measured. Note that it is necessary to use such geometrical distributions of the atoms (or molecules) and a test body, where oscillations (134) are "in phase" and, therefore, do not cancel each other.

Acknowledgments

We are thankful to Natalia N. Bagmet (Lebed), Vladimir A. Belinski, Steven Carlip, Fulvio Melia, Douglas Singleton, and Vladimir E. Zakharov for fruitful and useful discussions.

References

1. L. D. Landau and E. M. Lifshitz, *The Classical Theory of Fields*, 4th edn. Butterworth-Heinemann, Oxford (2003).
2. C. W. Misner, K. S. Thorne and J. A. Wheeler, *Gravitation*, Princeton University Press, Princeton (2017).
3. H. Pihan-le Barc, *et al.*, *Phys. Rev. Lett.* **123**, 231102 (2019).
4. P. Touboul, *et al.*, *Class. Quantum Grav.* **36**, 225006 (2019).
5. A. M. Nobili and A. Anselmi, *Phys. Rev. D.* **98**, 042002 (2018).
6. J. P. Pereira, J. M. Overduin and A. J. Poyneer, *Phys. Rev. Lett.* **117**, 071103 (2016).
7. A. G. Lebed, *Int. J. Mod. Phys. D.* **26**, 1730022 (2017).
8. A. G. Lebed, *J. Phys.: Conf. Ser.* **738**, 012036 (2016).
9. A. G. Lebed, *Int. J. Mod. Phys. D.* **28**, 1930020 (2019).
10. A. G. Lebed, *Mod. Phys. Lett. A.* **35**, 2030010 (2020).

11. D. Park, *Introduction to the Quantum Theory*, 3rd edn. Dover Publications, New York (2005).
12. K. Nordtvedt, *Class. Quantum Grav.* **11**, A119 (1994).
13. S. Carlip, *Am. J. Phys.* **66**, 409 (1998).
14. M. Zych, L. Rudnicki and I. Pikovski, *Phys. Rev. D.* **99**, 104029 (2019).
15. F. Schwabl, *Advanced Quantum Mechanics*, 3rd edn. Springer, Berlin (2005).
16. E. Fischbach, B. S. Freeman and W. K. Cheng, *Phys. Rev. D.* **23**, 2157 (1981).
17. Pierre Meystre, private communication, unpublished (2020).
18. N. D. Birrell and P. C. W. Davies, *Quantum Fields in Curved Space*, 3rd edn. Cambridge University Press, Cambridge (1982).

Chapter 2

Probing Unruh Radiation and the Equivalence Principle with Unruh–DeWitt Detectors

Emil T. Akhmedov[*,†,§], Joey Contreras[‡,¶] and Douglas Singleton[‡,∥]

[*]*Moscow Institute of Physics and Technology,*
Institutskii per. 9, 141700, Dolgoprudny, Russia
[†]*Institute for Theoretical and Experimental Physics,*
B. Cheremushkinskaya 25, 117218, Moscow, Russia
[‡]*Department of Physics, California State University Fresno,*
Fresno, CA 93740, USA
[§]*akhmedov@itep.ru*
[¶]*mkfetch@mail.fresnostate.edu*
[∥]*dougs@mail.fresnostate.edu*

Unruh–DeWitt detectors are quantum systems, which in their simplest form have two energy states, which can be used to probe whether or not particles are created by a given space-time background or by a particular path through a space-time. Unruh–DeWitt detectors can be used to probe phenomenon such as Unruh radiation (particle creation as seen by an accelerating observer), Hawking radiation (particle creation by a black hole) and Hawking–Gibbons radiation (particle creation in a de Sitter cosmological space-time). In this work, we first show that the natural Unruh–DeWitt detector of electrons in storage rings may already have experimentally confirmed the existence of Unruh radiation, but with the usual uniform linear acceleration replaced by uniform circular acceleration. In the lab frame, this effect is the already observed Sokolov–Ternov effect, while in the non-inertial frame of the electron the explanation is in terms of the Unruh effect. Next we show how the Unruh–DeWitt detector can be used to probe the equivalence principle. We compare the details of how the detector responses to Schwarzschild space-time, versus Rindler space-time (the space-time seen by a linearly accelerating observer). In both space-times, the Unruh–DeWitt detector registers particle, but the details are different. This allows one to distinguish, via a local measurement, between a gravitational field and acceleration, and thus leads to a violation of the equivalence principle.

1. Unruh–Dewitt Detector

To investigate, if a particular curved space-time (e.g. Schwarzschild space-time) or a particular space-time motion through flat, Minkowski space-time (e.g. uniform velocity, uniform, linear acceleration) exhibits radiation, we will use an Unruh–DeWitt detector.[1,2] A simple Unruh–DeWitt detector is a quantum system coupled to some field (which for simplicity we take to be a scalar field ϕ) with two energy states E_0 and E with $E_0 < E$. The coupling between the detector and field is given by the monopole interaction $g\mu(\tau)\phi(x(\tau))$ where g is the coupling constant, $\mu(\tau)$ is the detector's monopole moment, and $x(\tau) = x^\mu(\tau)$ is the detector's trajectory as a function of its proper time, τ. The transition rate per unit proper time, $T(E)$, for such a detector to be excited from its ground state E_0 to the higher energy E is given by (we set $G = c = 1$ here and throughout most of the paper unless otherwise noted)[3]

$$T(E) = g^2 \sum_E |\langle E|\mu(0)|E_0\rangle|^2 \int_{-\infty}^{+\infty} e^{-i(E-E_0)\Delta\tau} G^+(\Delta\tau) d(\Delta\tau). \quad (1)$$

A concrete realization of an Unruh–DeWitt detector is an electron in a magnetic fields as shown in Fig. 1. In this case, the low-energy state is when the magnetic moment of the electron aligns with the magnetic field and the spin therefore anti-aligns with the magnetic field. Also in this case the electron interacts not with a scalar field, ϕ, but with a vector field, A_μ. This difference is not important as it will only change the overall prefactor of $T(E)$ due to the vector field having more degrees of freedom as compared to the scalar field. Further details about the construction of an Unruh–DeWitt detector can be found in Refs. 3–6.

In the above expression $G^+(x, x') = \langle 0|\phi(x)\phi(x')|0\rangle$ is the positive frequency Wightman function. We use $G^+(x, x')$ since we are studying excitations from E_0 to E and $\Delta\tau = \tau - \tau'$. The negative frequency Wightman function is defined by $G^-(x, x') = \langle 0|\phi(x')\phi(x)|0\rangle$. The standard Feynman propagator, $G_F(x, x')$ is related to $G^\pm(x, x')$ via the time-ordered expression

$$iG_F(x, x') = \langle 0|\mathcal{T}[\phi(x)\phi(x')]|0\rangle$$
$$= \Theta(t - t')G^+(x, x') + \Theta(t' - t)G^-(x, x'), \quad (2)$$

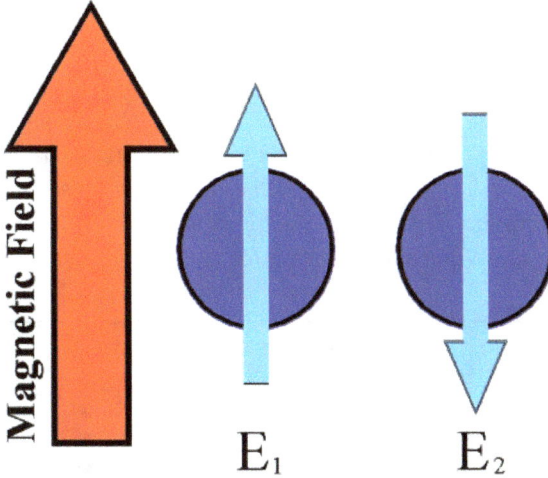

Fig. 1. A simple realization of an Unruh–DeWitt detector: an electron in a uniform magnetic field. This form of the Unruh–DeWitt detector will be used to discuss the experimental status of the observation of the Unruh effect via electrons in storage rings.

where $\mathcal{T}$ is the time-ordering operator and $\Theta(x)$ is the step function which equals 1 when $x > 0$ and 0 when $x < 0$.

For this work, the crucial part of (1) is the response function *per unit proper time* which is[3]

$$\mathcal{F}(E) = \int_{-\infty}^{+\infty} e^{-i(E-E_0)\Delta\tau} G^+(\Delta\tau) d(\Delta\tau). \qquad (3)$$

The Wightman function, G^+, is defined with respect to some vacuum state $|0\rangle$. Picking a different vacuum state can lead to different $\mathcal{F}(E)$'s. Some vacuum choices can lead to the detector being excited while other choices of vacuum leave the detector in the ground state. This leads to the apparently contradictory statement that whether or not an Unruh–DeWitt detector will get excited for a given space-time metric or space-time path is observer dependent.[3,7,8] However, whether or not there are particles in a given space-time background or along a given space-time path, and whether or not an Unruh–DeWitt detector gets excited, is observer dependent. From (3) one can see that the response function, $\mathcal{F}(E)$, depends only on the space-time trajectory of the detector and is independent of its coupling and monopole moment.

2. Unruh Effect

In this section, we use (1), with the Unruh–DeWitt detector initially in its ground state, (i.e. E_0), but with the detector moving along some given trajectory. We want to determine if the detector has a non-zero excitation rate (i.e. if $T_- \neq 0$) as it moves along a trajectory in the background QFT, which we take to be (for simplicity) one massless, real scalar field, ϕ with a monopole coupling to the detector given by $g\mu(\tau)\phi(x(\tau))$. Under these conditions the 4D Wightman function for the massless scalar field is

$$G_+(\tau) \propto \frac{1}{|x_0(t+\tau/2) - x_0(t-\tau/2) - i\,\epsilon|^2 - |\vec{x}(t+\tau/2) - \vec{x}(t-\tau/2)|^2}.$$

We have replaced $\Delta\tau$ by simply $\tau = t - t'$. From (1) the transition rate per unit time becomes

$$T_\mp \propto \lim_{\epsilon \to 0} g^2 \, |\langle 0\,|\mu(0)|\,1\rangle|^2 \int_{-\infty}^{+\infty} d\tau \tag{4}$$

$$\times \frac{e^{\mp i\,\Delta E\,\tau}}{|x_0(t+\tau/2) - x_0(t-\tau/2) - i\,\epsilon|^2 - |\vec{x}(t+\tau/2) - \vec{x}(t-\tau/2)|^2},$$

with $\Delta E = E - E_0$. To obtain a specific transition rate, $T_\mp$, we now need to substitute into (4) whichever trajectory, $x(t)$, we are interested in. In general $T_\mp$ explicitly depends on time t. However, for homogeneous motions this dependence on time disappears, and this greatly simplifies the calculation of the integral in (4).

2.1. Motion with constant velocity

For our first example, we look at motion with a constant velocity. In the case with $v = $ const. the trajectory is given by $x(t) = (\gamma t, \gamma v t, 0, 0)$, with $\gamma = 1/\sqrt{1 - v^2}$ and t is the detector's proper time. Inserting this trajectory into (4), we obtain

$$T_\mp \propto \lim_{\epsilon \to 0} g^2 \, |\langle 0\,|\mu(0)|\,1\rangle|^2 \int_{-\infty}^{+\infty} d\tau \, \frac{e^{\mp i\,\Delta E\,\tau}}{(\gamma\,\tau - i\,\epsilon)^2 - (\gamma\,v\,\tau)^2}. \tag{5}$$

$T_\mp$ does not depend on t since the motion is homogeneous. If instead of considering the phenomenon from the point of view of the comoving reference frame we had considered the laboratory reference frame, $T_\mp$ would be changed by a factor of γ because of time dilation. In (5) we are using the standard regularization of the Wightman function which shifts the pole at $\tau = 0$ to the upper complex half-plane. As we will see in a moment such a regularization is necessary to have a non-zero probability for the spontaneous radiation of the detector if it is originally in the excited state.

The integral in (5) is taken using contour integration in the complex τ plane. Since $E > E_0$, the integral in (5), for T_- uses a contour which is closed with a large semi-circle in the lower complex half-plane. This contour is denoted by C_-. For T_+ the contour is closed with a large semi-circle in the upper complex half-plane, and is denoted by C_+. It is this choice of the contours which we will use for $T_\mp$ everywhere below.

The expression in the denominator of the integrand in (5) has two zeros located at

$$\tau_\pm = i\epsilon \sqrt{\frac{1 \pm v}{1 \mp v}}. \tag{6}$$

Hence, the transition rate per unit time is

$$T_\mp \propto \lim_{\epsilon \to 0} g^2 \, |\langle 0 \,|\mu(0)|\, 1\rangle|^2 \oint_{C_\mp} d\tau \, \frac{e^{\mp i \Delta \mathcal{E} \tau}}{\tau_+ - \tau_-} \left(\frac{1}{\tau - \tau_+} - \frac{1}{\tau - \tau_-} \right). \tag{7}$$

Evaluating the integrals gives

$$T_- = 0, \quad T_+ \propto g^2 \, |\langle 0 \,|\mu(0)|\, 1\rangle|^2 \, \Delta E. \tag{8}$$

$T_- = 0$ because C_- does not enclose any poles, while $T_+ \neq 0$ since C_+ does enclose poles. This result in (8) is independent of the velocity and, hence, is also valid for the static detector with $v = 0$. The physical meaning of the results in (8) is the following: For a detector moving with constant velocity in the vacuum of a QFT, there is zero probability for the detector to get excited i.e. $T_- = 0$. But, if a detector starts off in the excited state, there is a non-zero probability for it to radiate spontaneously i.e. $T_+ \propto \Delta E$.

2.2. *Motion with constant linear acceleration*

We next turn to the case of constant, linear acceleration a. This time the space-time position is given by $x(t) = \left(\frac{1}{a} \sinh [a\, t], \frac{1}{a} \cosh [a\, t], 0, 0 \right)$, where again t is the detector's proper time. Substituting this expression into (4) gives

$$T_\mp \propto \lim_{\epsilon \to 0} g^2 \, |\langle 0 \,|\mu|\, 1\rangle|^2 \oint_{C_\mp} d\tau \, e^{\mp i \Delta E \tau} \, \frac{a^2}{\sinh^2 \left[\frac{a}{2} \left(\tau - i\,\epsilon \right) \right]}. \tag{9}$$

Again the rate $T_\mp$ does not depend on the time because the motion is homogeneous, and again in (9) the standard regularization of the Wightman function is used.[9]

The expression in (9) is essentially the same as a stationary detector in contact with a heat bath at temperature T. For a heat bath the transition rate is given by[4]

$$T_{\text{heat bath}} \propto \lim_{\epsilon \to 0} g^2 \, |\langle 0 \, |\mu(0)| \, 1 \rangle|^2 \int_{-\infty}^{+\infty} d\tau \, e^{\mp i \Delta E \tau} \, \frac{T^2}{\sinh^2 [\pi T \, (\tau - i \epsilon)]}, \tag{10}$$

On comparing (9) with (10) one finds that a detector moving with constant, linear acceleration in the background QFT (which is originally in the vacuum state) will get excited. Additionally, the accelerating detector detects particles with the standard Planckian/thermal distribution of a heat bath. The temperature of the accelerating detector can be read off by comparing (9) and (10) and is as follows[1]:

$$T = \frac{a}{2 \pi}. \tag{11}$$

If we restore unit to (11) we would have $T = \frac{a \hbar}{2 k_B c \pi}$. In order to achieve a temperature of 1 Kelvin, one would need an acceleration of $a \approx 10^{20} \frac{m}{s^2}$ which is enormous. Such accelerations would completely destroy any finite, sized, macroscopic object. Furthermore, linear accelerations of this size would mean that the detector would very quickly be accelerated beyond the spatial domain of an observer who would want to make a measurement with this detector. For these reasons, we will turn below to electrons in storage rings. In storage rings, one can get centripetal accelerations on the order of $a \approx 10^{23}$–$10^{24} \frac{m}{s^2}$ implying, via (11), temperatures in 10^3–10^4 kelvin. Furthermore, electrons are point-like objects and thus are able to withstand these enormous accelerations without being distorted or destroyed.

From Eq. (10), one can see that a static detector in a heat bath gets excited due to the absorption of thermal excitations from the background QFT. This raises the question as to why the accelerating detector becomes excited if it is in a vacuum state of the QFT? The general physical explanation is that both the detector and QFT are excited at the cost of work performed by the force driving the detector along its trajectory. In greater detail this can be explained as follows: The Hamiltonian of the background QFT in the detector's comoving, non-inertial reference frame has negative energy eigenstates.[10] The detector can radiate these negative energy eigenstates which then excites the detector. The energy for this comes from the external driving force, i.e. the system is not closed. We can explain the negative energy eigenstates by looking at the Wightman function which

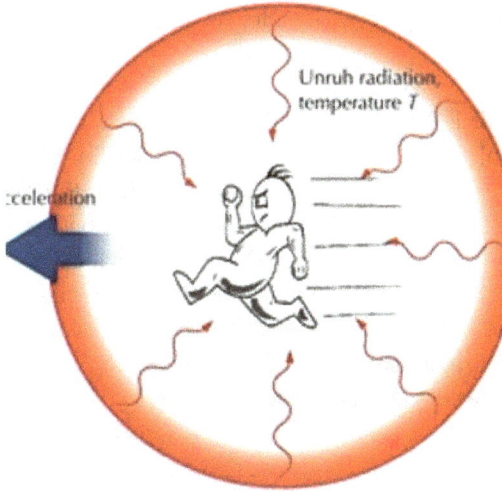

Fig. 2. A cartoon picture of the canonical Unruh effect: a linearly accelerating observer will detect radiation with a thermal spectrum.

was used in the derivation of (9) or (5) under different types of motion:

$$G[x(\tau),\ x(0)] \propto \frac{1}{|x(\tau) - x(0)|^2}$$

$$\propto \int \frac{d^3 p}{4\,\pi^2} \frac{1}{\omega} \exp\left\{-i\,p\left[x(\tau) - x(0)\right]\right\}\big|_{\omega=|\vec{p}|}. \qquad (12)$$

First consider constant velocity motion where $x(t) = (\gamma\,t, \gamma\,\vec{v}\,t)$. Plugging this representation of the Wightman function with this trajectory $x(t)$ into (4), we obtain

$$T_\mp \propto g^2\ |\langle 0\,|\mu(0)|\,1\rangle|^2 \qquad (13)$$

$$\times \int_{-\infty}^{+\infty} d\tau \int \frac{d^3 p}{4\,\pi^2} \frac{1}{\omega} \exp\left\{-i\left[\pm\Delta E + \gamma\left(\omega - \vec{p}\cdot\vec{v}\right)\right]\tau\right\}\big|_{\omega=|\vec{p}|}.$$

Taking the integral over τ first gives an energy conserving δ-function as the integrand of the $d^3 p$ integration

$$\delta\left[\pm\Delta E + \gamma\left(\omega - \vec{p}\cdot\vec{v}\right)\right]. \qquad (14)$$

This function vanishes for the upper (+) sign, in its argument. For the + sign, the argument is always greater than zero, because $E > E_0$ and

$|\vec{p}| = \omega > \vec{p} \cdot \vec{v}$, because $v < 1$.[3] Hence, in the case of constant velocity motion $T_- = 0$, but $T_+ \neq 0$, which follows just from energy conservation.

If the detector moves in an environment where its velocity is greater than the speed of light, it can produce Cherenkov-type radiation. In this case, it is possible to see that now the δ-function in (14) can be non-zero even for the case when the sign in front of ΔE is the "+" sign, i.e. the argument in (14) can equal zero for some angles between $\vec{p}$ and $\vec{v}$. This demonstrates the anomalous Doppler effect,[11] which describes the well-understood phenomenon that in such circumstances the detector can both radiate and have its internal degrees of freedom excited. A similar thing happens in the case of accelerated motion.

In the linear acceleration case we find

$$
T_\mp \propto g^2 \, |\langle 0 \, |\mu(0)| \, 1 \rangle|^2 \int_{-\infty}^{+\infty} d\tau \int \frac{d^3 p}{4 \pi^2} \frac{1}{\omega}
$$

$$
\times \exp \left\{ -i \left(\pm \Delta E \tau + \frac{\omega}{a} \sinh(a\tau) - \frac{\vec{a} \cdot \vec{p}}{a^2} \left[\cosh(a\tau) - 1 \right] \right) \right\} \bigg|_{\omega = |\vec{p}|}.
\tag{15}
$$

After the integration over τ one will *not* obtain zero regardless of the sign chosen for ΔE. Thus, in this case, energy conservation allows the detector to absorb positive energy states or, equivalently radiate negative energy states. To see this explicitly one has to take the integral over τ in the saddle-point approximation (if $\Delta E/a \gg 1$) and include contributions of all saddle points of τ, which are related to the pole contributions in (9). A clearer physical picture for the appearance of the radiation in the non-inertial reference frame can be obtained in the quasi-classical quantization scheme for relativistic particles in curved stationary backgrounds.[4, 5] In the latter case, it is straightforward to see that the radiation appears due to polarization of the vacuum in strong gravitational background fields (see Ref. 12 for a more detailed discussion on this issue).

All well and good, but the case of constant homogeneous linear acceleration is not possible to arrange in reality — one cannot have an eternally accelerating detector. If one does consider a more realistic motion (e.g. a stationary detector which accelerates for a finite time and then moves with constant velocity) the initial and final conditions increase the difficulty of the analysis making it much harder (or impossible) to get a clear physical picture of what is going on. In particular it is not clear whether or not there will be a non-trivial saddle-point contribution as in (15) if the acceleration is over a finite time.

Moreover, we have considered the phenomenon in question from the point of view of non-inertial, comoving reference system. If instead we study the phenomenon from the point of view of the laboratory, inertial reference frame, then the trajectory is $x(t) = (t, \sqrt{1/a^2 + t^2}, 0, 0)$. Now the motion does not look homogeneous and $T_\mp$ seems to explicitly depend on time, which again makes the study of the phenomenon difficult.

For these reasons we turn our attention to circular motion. We will consider homogeneous circular motion, i.e. eternal circular motion with no starting or stopping. However, we will show that homogeneous circular motion is a good approximation for real circular motion with a starting and stopping time. In the section, following we will discuss under what conditions one could observe the circular Unruh effect using charged particles with spin in a magnetic field as the Unruh–Dewitt detectors.

2.3. *Motion with constant circular acceleration*

If the detector performs homogeneous, circular motion[14] with radius R and with the angular velocity ω_0, then the trajectory is given by

$$x(t) = (\gamma t,\, R \cos [\gamma \omega_0 t],\, R \sin [\gamma \omega_0 t],\, 0),\, \gamma = 1/\sqrt{1 - R^2 \omega_0^2}$$

and t is the detector's proper time.

Inserting this trajectory into (4), we obtain

$$T_\mp \propto \lim_{\epsilon \to 0} g^2 \, |\langle 0 \,|\mu(0)|\, 1\rangle|^2 \oint_{C_\mp} d\tau \, \frac{e^{\mp i \,\Delta E \,\tau}}{[\gamma \,(\tau - i\,\epsilon)]^2 - 4\,R^2 \sin^2 \left[\frac{\gamma \,\omega_0}{2}\,\tau\right]}. \tag{16}$$

Again $T_\mp$ does not depend on t, because the motion is homogeneous. Also the integration is over the proper time, but since circular motion has a simple relationship between proper time, τ, and laboratory time ($\tau_L = \gamma \,\tau$) one can easily change to the laboratory frame as in the case of the motion with constant velocity. This is an important difference between circular and linear acceleration which makes the analysis of the circular case more straightforward. Furthermore, in the case of circular motion one can easily transform the expression in (16) from the laboratory reference frame to either the inertial *instantaneously*, comoving reference frame or the non-inertial, comoving reference frame. Such transformations cannot be so easily done in the case of linearly accelerating motion.

The integrand in (16) has poles in the complex τ plane. In particular, it has poles similar in nature to those in the Wightman function for a heat bath or for linear acceleration, which lead to a Boltzmann-type exponential contribution to $T_{\mp}$. However, there are also pre-exponential contributions to $T_{\mp}$ for circular motion which spoil the exact thermal behavior.

Following Ref. 15, let us compare the Wightman functions for the cases of linear and circular acceleration:

$$G_L(\tau) \propto \frac{a^2}{\sinh^2\left[\frac{a}{2}\left(\tau - i\,\epsilon\right)\right]}$$

$$\approx \frac{1}{(\tau - i\,\epsilon)^2}\left(1 + \frac{1}{12}\,[a\,\tau]^2 + \frac{1}{360}\,[a\,\tau]^4 + \cdots\right)^{-1},$$

$$G_C(\tau) \propto \frac{1}{[\gamma\,(\tau - i\,\epsilon)]^2 - 4\,R^2\,\sin^2\left[\frac{\gamma\,\omega_0}{2}\,\tau\right]}$$

$$\approx \frac{1}{(\tau - i\,\epsilon)^2}\left(1 + \frac{1}{12}\,[a\,\tau]^2 - \frac{1}{360\,v^2}\,[a\,\tau]^4 + \cdots\right)^{-1}, \quad (17)$$

where for the case of circular motion the velocity is $v = \omega_0\,R\,\gamma$ and the acceleration is $a = \gamma^2\,\omega_0^2\,R$ in the comoving frame. Note that the difference between $G_L(\tau)$ and $G_C(\tau)$ appears only in the third term on the RHS of both equations.

The integral in (16) cannot be done exactly in contrast to the linear acceleration case (see Ref. 16 for a semi-analytical study of this integral in various limiting regimes). However, as we show in the following section the circular case simplifies for large γ. Assuming that the energy splitting is not too small (i.e. $\Delta E > a$) we can approximate the Wightman function by

$$G_C(\tau) \approx \frac{1}{(\tau - i\,\epsilon)^2}\left(1 + \frac{1}{12}\,[a\,\tau]^2\right)^{-1}. \quad (18)$$

The integral in (16) can now be computed with the result

$$T_- \propto g^2\,|\langle 0\,|\mu(0)|\,1\rangle|^2\,a\,e^{-\sqrt{12}\,\frac{\Delta E}{a}},$$

$$T_+ \propto g^2\,|\langle 0\,|\mu(0)|\,1\rangle|^2\,a\left(e^{-\sqrt{12}\,\frac{\Delta E}{a}} + 4\,\sqrt{3}\,\frac{\Delta E}{a}\right). \quad (19)$$

The exponential contribution comes from the non-trivial pole in (18) at $\tau = \pm i\,2\sqrt{3}/a$. The second contribution to T_+ comes from the trivial pole at $\tau = 0$.

For $\Delta E \gg a$, the equilibrium population of the upper level relative to the lower level is

$$\mathcal{P}_0 = \frac{T_+ - T_-}{T_+ + T_-} \approx 1 - \frac{1}{4\sqrt{3}} \frac{a}{\Delta E} e^{-2\sqrt{3}\frac{\Delta E}{a}}. \tag{20}$$

This is exactly the kind of equilibrium "polarization" we will obtain in the following section when we study the Sokolov–Ternov effect for particles with large gyromagnetic number g. Note that this equilibrium "polarization" is not thermal due to the dependence of the pre-exponential factor on $\frac{a}{\Delta E}$. Furthermore, even if we take into account corrections to (20) we would not expect to get a thermal spectrum of the detected particles. Intuition from condensed matter informs us that the Planckian distribution is strongly related to the form of the two-point correlation function in (9). The two-point function for circular motion given in (16) has a drastically different form than that in (9).

The explanation why both T_+ and T_- are not zero in this case is the same as the one given at the end of the previous subsection which dealt with linear acceleration. Here we find:

$$T_\mp \propto g^2 \, |\langle 0 \,|\mu(0)|\, 1\rangle|^2$$

$$\times \int_{-\infty}^{+\infty} d\tau \int \frac{d^3 p}{4\pi^2} \frac{1}{\omega} \exp\{-i[(\pm\Delta\mathcal{E} + \gamma\omega)\tau - \vec{p}\cdot\vec{R}(\tau)]\}|_{\omega=|\vec{p}|}, \tag{21}$$

where $\vec{R}(\tau) = (R\cos[\gamma\omega_0\tau] - 1, R\sin[\gamma\omega_0\tau], 0)$. Again after taking the integral over τ the resulting expression under the integral over $d^3 p$ is not zero for any choice of sign in front of ΔE, which means that the detector can emit negative energy states, i.e. get excited to E. Again quasi-classical approxiamation gives a clearer picture of the phenomenon. According to Ref. 10 (see Refs. 4 and 5 for the quasi-classical study), the detector clicks due to falling particles to the center (orbiting detector) from the vacuum fluctuations. See as well Ref. 17 for a more detailed discussion of this issue.

3. Sokolov–Ternov Effect with Arbitrary Gyromagnetic Number and Unruh Effect

In this section, we will use an electrically charged particle with spin which is undergoing circular motion as our detector for the circular Unruh effect.

We now give the conditions under which the particle's motion can be considered classical. We will mostly be discussing electrons, but we keep the gyromagnetic number g arbitrary. Unlike the previous cases, in this section we will only consider the laboratory reference frame and laboratory time.

There are two sources of quantum effects in synchrotron radiation: (i) the quantization of the electron's trajectory and (ii) quantum back-reaction under photon emission. The first one is suppressed if $\omega_0/\mathcal{E} \ll 1$, where $\mathcal{E}$ is the electron's energy and $\omega_0 = e H/\mathcal{E}$ is the angular velocity of the electron moving in a background magnetic field H, i.e. ω_0 is the energy splitting between the Landau levels.

The second source of quantization is defined by the ratio $\omega_c/\mathcal{E}$, where

$$\omega_c = \omega_0 \left(\frac{\mathcal{E}}{m}\right)^3 = \omega_0\, \gamma^3, \qquad (22)$$

is the characteristic frequency of the photons emitted in the synchrotron radiation.[19, 20] Here m is the mass of the electron. If the ratio in question satisfies

$$\frac{\omega_c}{\mathcal{E}} \ll 1 \qquad (23)$$

then the electron is ultra-relativistic and its motion is essentially classical.

Another approximation which we adopt comes from the characteristic features of the radiation in the ultra-relativistic case. Consider a relativistic electron with $\gamma = \frac{\mathcal{E}}{m} \gg 1$. The angular distribution of the radiated power in this ultra-relativistic limit is proportional to large powers of

$$\frac{1}{(1 - \vec{n} \cdot \vec{v})} \qquad (24)$$

where $\vec{n} = \vec{p}/\omega$ and $\vec{p}$ and ω are the photon's momentum and energy.[20, 21] Because of the large negative powers of $1 - \vec{n} \cdot \vec{v}$, the radiation is concentrated in a narrow cone around the direction of the velocity $\vec{v}$. The angle of the cone is approximately defined by ($v \approx 1$)

$$1 - \vec{n} \cdot \vec{v} = 1 - v \cos\theta \approx 1 - v + \frac{\theta^2}{2} \approx \frac{1}{2}\left(\frac{1}{\gamma^2} + \theta^2\right), \qquad (25)$$

where θ is the angle between the velocity $\vec{v}$ and the radiation direction $\vec{n}$. Hence, the angle of the radiation cone is

$$\theta < \frac{1}{\gamma} = \frac{m}{\mathcal{E}}. \qquad (26)$$

Thus, the radiation in a given direction is formed from the small part of the trajectory, over which the velocity vector $\vec{v}$ is rotated by the angle $m/E \ll 1$. The electron covers this part of the trajectory in a laboratory time Δt given by

$$\Delta t \, |\dot{\vec{v}}| \approx \Delta t \, \omega_0 \leq \frac{1}{\gamma} \ll 1. \tag{27}$$

It is this interval of time which gives the main contribution to the integrals we calculate below.

3.1. *Synchrotron radiation due to the electric charge*

There are two ways the electron radiates. The first one is the well-known synchrotron radiation of a charged particle. The interaction Hamiltonian in this case is as follows:

$$\hat{H}_{int} = e \, \vec{A} \cdot \hat{\vec{v}}, \tag{28}$$

where the velocity operator is $\hat{\vec{v}} = (-i/m) \, \nabla$ and the vector potential $\vec{A}$ is taken to be in the radiation gauge, $\nabla \cdot \vec{A} = 0$. The vector potential of a

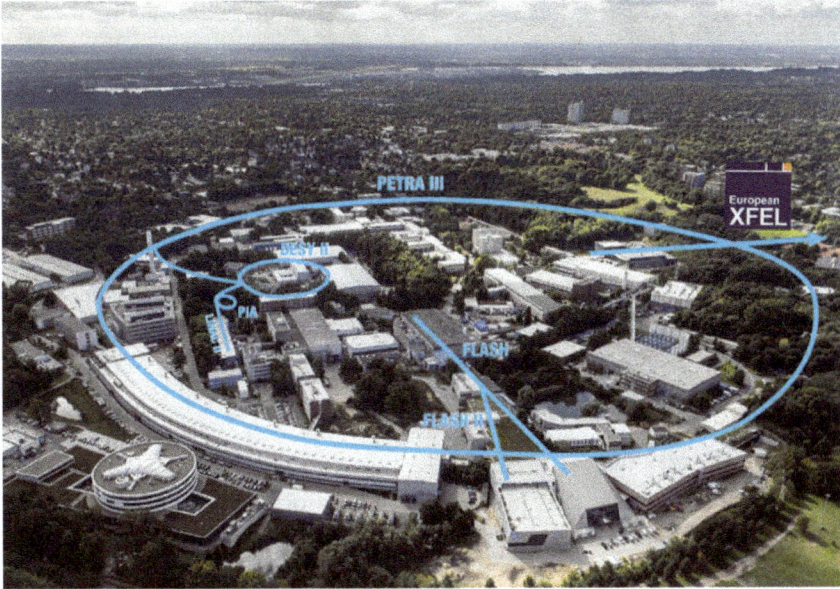

Fig. 3.   The Sokolov–Ternov effect — the partial polarization of electrons in storage rings — has been observed in facilities like DESY.

plane electromagnetic wave with the polarization $\vec{\zeta}$ is

$$\vec{A}(\vec{r},\,t) = \vec{\zeta}\,\sqrt{\frac{2\,\pi}{\omega}}\,e^{-\mathrm{i}\,[\omega\,t-\vec{p}\cdot\vec{r}]} + \text{c.c.,} \qquad \text{where } \omega = |\vec{p}|. \qquad (29)$$

The radiation probability of a photon with the momentum in the range $[\vec{p},\,\vec{p}+d\vec{p}]$ is equal to

$$dP = \left|\,\mathrm{i} \int_{-\infty}^{+\infty} \left\langle \Psi_f(t) \left| \hat{H}_{int}(t) \right| \Psi_i(t) \right\rangle dt \right|^2 \frac{d^3p}{(2\,\pi)^3}, \qquad (30)$$

where $\vec{p}$ is the photon momentum. The integral over the laboratory time t should be taken over the period of the circular motion.[19,20] However, we have taken the integration region over the whole real line, because the integral is saturated in the saddle-point approximation by the small region of t given in (27).

As in the introduction, we change the integration variables to $\tau = t - t'$ and $\tau' = t + t'$ and drop the integral over τ'. In this way, we obtain an expression for dw — the radiation rate per unit time and per infinitesimal momentum $d\vec{p}$. The infinitesimal radiation power is $dI_o = \omega\,dw$. Taking the interaction Hamiltonian which corresponds to photon emission gives

$$dI_o = \frac{e^2\,d^3p}{4\,\pi^2} \int_{-\infty}^{+\infty} d\tau$$

$$\cdot \left\langle \Psi_i\left(t+\frac{\tau}{2}\right) \left| \vec{\zeta}\cdot\hat{\vec{v}}\left(t+\frac{\tau}{2}\right) e^{-\mathrm{i}\left[\omega\left(t+\frac{\tau}{2}\right)-\vec{p}\cdot\hat{\vec{r}}\left(t+\frac{\tau}{2}\right)\right]} \right| \Psi_f\left(t+\frac{\tau}{2}\right) \right\rangle$$

$$\cdot \left\langle \Psi_f\left(t-\frac{\tau}{2}\right) \left| \vec{\zeta}^*\cdot\hat{\vec{v}}\left(t-\frac{\tau}{2}\right) e^{\mathrm{i}\left[\omega\left(t-\frac{\tau}{2}\right)-\vec{p}\cdot\hat{\vec{r}}\left(t-\frac{\tau}{2}\right)\right]} \right| \Psi_i\left(t-\frac{\tau}{2}\right) \right\rangle,$$

$$(31)$$

where the velocity $\hat{\vec{v}}(t)$ and the coordinate $\hat{\vec{r}}(t)$ are the Heisenberg operators. In the quasi-classical approximation adopted here, these operators can be substituted by their classical values, i.e.

$$\langle \Psi_f(t)| \vec{\zeta}^* \cdot \hat{\vec{v}}(t)\, e^{\mathrm{i}\,[\omega\,t-\vec{p}\cdot\hat{\vec{r}}(t)]}|\Psi_i(t)\rangle \to \vec{\zeta}^* \cdot \vec{v}(t)\, e^{\mathrm{i}\,[\omega\,t-\vec{p}\cdot\vec{r}(t)]}, \qquad (32)$$

where $\vec{v}(t) = \dot{\vec{r}}(t)$ and $\vec{r}(t) = (r_1(t),\,r_2(t),\,r_3(t))$ are now classical velocities and coordinates along the trajectory in laboratory time t:

$$r_1(t) = x_0 + \frac{m}{e\,H}\,\sin\left(\frac{e\,H}{\mathcal{E}}\,t+\varphi\right),$$

$$r_2(t) = y_0 + \frac{m}{e\,H}\,\cos\left(\frac{e\,H}{\mathcal{E}}\,t+\varphi\right). \qquad (33)$$

These equations for the trajectory are correct if the electron is ultra-relativistic ($|\vec{v}| \approx 1$) and moving in the plane perpendicular to the magnetic field $\vec{H}$. The initial conditions are given by x_0, y_0, φ. Substituting (33) into (32) and then into (31) and summing over the photon polarizations and integrating over $d\tau$ in the saddle-point approximation yields, in the large ω limit[19]:

$$\frac{dI_o}{d\omega} \approx \frac{1}{2\sqrt{\pi}} \frac{e^3 \omega_c}{\gamma^2} \left(\frac{\omega}{\omega_c}\right)^{\frac{1}{2}} \exp\left\{-\frac{2\omega}{3\omega_c}\right\}, \quad \text{where } \omega_c = \omega_0 \gamma^3. \tag{34}$$

This equation shows that the main contribution to the radiation comes from the photons with frequency around ω_c. This confirms the discussion around (22). If in (31) we had instead taken the integrals over both $\vec{p}$ and τ in the appropriate approximation, we would have obtained

$$I_0 = \frac{2e^2}{3R} \gamma^4 \omega_0 \,,$$

which is the well-known total formula for power radiated by a charge undergoing uniform circular motion in the quasi-classical approximation.

3.2. *Synchrotron radiation due to spin flip*

In addition to the radiation associated with its charge, the electron can also radiate via a spin flip transition. The energy distribution of this radiation is similar to the one given in (34).[18] We are interested in the probability rate of the radiation, which can be obtained from the relativistic motion of a spin $\vec{s}$ as given by[19]

$$\frac{d\vec{s}}{dt} = i\left[\hat{H}_{int}, \vec{s}\right], \tag{35}$$

$$\hat{H}_{int} = -\frac{e}{m} \vec{s} \left[\left(\alpha + \frac{1}{\gamma}\right) \vec{H} - \frac{\alpha\gamma}{\gamma+1} \vec{v} \left(\vec{v} \cdot \vec{H}\right) - \left(\alpha + \frac{1}{\gamma+1}\right) \vec{v} \times \vec{E}\right],$$

where $\alpha = (g-2)/2$ is the magnetic moment anomaly, $\vec{v}$ is the particles velocity and $\vec{E}$ is the electric field. Substituting the vector potential for the outgoing photon from (29) into (36) and then substituting this Hamiltonian into (30) yields the total probability rate (in the quasi-classical approximation of the electron's motion)

$$T_{\mp} = \frac{e^2}{4\,\pi^2\,m^2} \sum_{\text{phot. pol.}} \int \frac{d^3p}{\omega} \int_{-\infty}^{+\infty} d\tau$$

$$\cdot \left\langle i \left| s_k^* \left(t + \frac{\tau}{2} \right) \right| f \right\rangle \left\langle f \left| s_j \left(t - \frac{\tau}{2} \right) \right| i \right\rangle$$

$$\cdot W_{km} \left(t - \frac{\tau}{2} \right) \zeta_m^* \, W_{lj} \left(t + \frac{\tau}{2} \right) \zeta_l$$

$$\cdot \exp \left\{ -\mathrm{i} \left(\omega\tau - \vec{p} \cdot \left[\vec{r} \left(t - \frac{\tau}{2} \right) - \vec{r} \left(t + \frac{\tau}{2} \right) \right] \right) \right\}, \qquad (36)$$

where we have taken the sum over the photon polarizations and the W_{il} are given by

$$W_{il}(\vec{p}, \, \omega, \, t) = \left[\left(\alpha + \frac{1}{\gamma} \right) \epsilon_{ijl}\, p_j - \frac{\alpha\,\gamma}{\gamma+1}\, v_i(t)\epsilon_{jml}\, v_j(t)\, p_m \right.$$

$$\left. - \left(\alpha + \frac{1}{\gamma+1} \right) \epsilon_{ijl}\, v_j(t)\, \omega \right] \qquad (37)$$

and $\vec{v}(t) = \dot{\vec{r}}(t)$. The origin of the subscripts $\mp$ in the LHS of (36) will be explained in a moment.

In (36), the integral over τ should be over the period of the circular motion. However, we can extrapolate the integration region to the whole real line, because the integral is saturated by the small region of values of τ defined in (27). It is this approximation which allows us to model the realistic motion of charged particles in storage rings by homogeneous circular motion.

To evaluate $T_{\mp}(t)$ further we need to evaluate the expectation values of the spin operators. From (36) with a constant magnetic field, H, the spin operators evolve in laboratory time according to

$$s_{\pm}(t) = s_1(t) \pm \mathrm{i}\, s_2(t) = s_{\pm}(0)\, e^{\pm \mathrm{i}\,\omega_s\, t}, \quad s_3(t) = s_3(0),$$

$$\omega_s = \left[1 + \gamma \left(\frac{g-2}{2} \right) \right] \omega_0, \qquad (38)$$

where ω_s is the precession frequency of the spin in the external, constant magnetic field, i.e. it is the energy difference $E - E_0$ between the upper and lower spin energy levels. Because we are interested in spin flip transitions we take the initial value of the spin ($|i\rangle$) either along or against the magnetic field and then flip it ($\langle f| \neq \langle i|$). This yields

$$\langle f\,|s_+(0)|\,i\rangle = \left\langle f\left|\frac{1}{4}\,(\sigma_1 + i\sigma_2)\right|i\right\rangle = -\frac{1}{4}\,(1 \mp 1),$$

$$\langle f\,|s_-(0)|\,i\rangle = \left\langle f\left|\frac{1}{4}\,(\sigma_1 - i\sigma_2)\right|i\right\rangle = \frac{1}{4}\,(1 \pm 1),$$

$$\langle f\,|s_z(0)|\,i\rangle = \left\langle f\left|\frac{1}{2}\sigma_3\right|i\right\rangle = 0. \tag{39}$$

In the first two equations, the upper sign corresponds to the spin flip which increases the spin energy, while lower sign decreases the spin energy. These signs correspond to the signs in (36). In addition using sum over photon polarizations

$$\sum_{\text{phot. pol.}} \zeta_m^* \, \zeta_l = \delta_{ml} - \frac{p_m\,p_l}{p^2}, \tag{40}$$

and

$$G(\vec{r}, t) = \frac{4\pi}{(t - i\epsilon)^2 - \vec{r}^2} = \int \frac{d^3\vec{p}}{\omega}\,\exp\{-i\,(\omega\,t - \vec{p}\cdot\vec{r})\}\Big|_{\omega = |\vec{p}|}, \tag{41}$$

the probability rates are

$$T_\mp(t) = \frac{e^2}{4\pi^2\,m^2}\,\lim_{\epsilon\to 0}\,\left\langle i|s_k^*(0)\,|f\right\rangle \langle f|s_j(0)\,|i\rangle \oint_{C_\mp} d\tau$$

$$\times \left[\hat{W}_{km}\left(t - \frac{\tau}{2}\right)\,\hat{W}_{mj}\left(t + \frac{\tau}{2}\right) + \left(\alpha + \frac{1}{\gamma+1}\right)^2\right.$$

$$\times \left. \epsilon_{knm}\,v_n\,\epsilon_{jli}\,v_l\,\frac{\partial}{\partial r_m}\,\frac{\partial}{\partial r_i}\right]$$

$$\times \frac{4\pi\,e^{\mp i\omega_s\,\tau}}{(\tau - i\epsilon)^2 - [\vec{r} - \vec{r}']^2}\Bigg|_{r=r(t-\frac{\tau}{2}),\ r'=r(t+\frac{\tau}{2})}, \tag{42}$$

where $\hat{W}$ is the differential operator obtained by substituting the operators $\vec{p} = i\partial/\partial\vec{r}$ and $\omega = i\partial/\partial t$ into (37). Taking a homogeneous circular trajectory for $r(t)$, we find that, up to the pre-exponential differential operator, (42) coincides with (16). The energy difference $E - E_0$ is replaced by ω_s, since now this energy difference comes from the electron's spin in a constant, background magnetic field. This pre-exponential is the source of the difference between the standard Sokolov–Ternov effect (where the detector is coupled to the electromagnetic field) and the circular Unruh effect (where the detector is coupled to a scalar field). Usually one takes the Unruh effect as being given only by the exponential contribution to $T_\mp$ as in (19).

To take the integral in (42) we change the integration variable to $z = \tau\,\omega_0\,\gamma$, take the contours, $C_\mp$, defined in the previously, use the approximate expression for the Wightman function (18) and use the standard integrals

$$
\lim_{\epsilon \to 0} \oint_{C_-} \frac{e^{-i A z}\, dz}{(z - i\,\epsilon)^n \left(1 + \frac{z^2}{12}\right)^m}
$$

$$
= \frac{i^n\, e^{-A\sqrt{12}}\, \pi \left(\sqrt{12}\right)^{1-n}}{(m-1)!}
$$

$$
\times \left(\frac{n+1}{2}\right) \left(\frac{n+1}{2} + 1\right) \cdots \left(\frac{n+1}{2} + m - 2\right), \quad m \geq 1, \quad (43)
$$

and similarly for C_+. The non-trivial pole contribution in (42) from the integral over τ is the same as the saddle-point contribution which plays its role if in (36) one takes the integral over τ first and then takes the integral $d^3 p$. An important point to note is that after substituting the contribution of the non-trivial pole ($z = \pm i\, 2\sqrt{3}$) into the exponent of (42) we obtain an expression proportional to $e^{-(1/\gamma + \sqrt{12}\,\alpha)}$. If $g \approx 2$ (the case of electrons) and $\gamma \gg 1$ the exponential factor ≈ 1. This is the reason why its contribution is usually overlooked in the standard Sokolov–Ternov calculation.[13]

Combining (42), (18), (43) and considering only $\alpha > 0$ yields[18]

$$
T_\mp \approx \frac{1}{2\,\tau_0} \{F_1(\alpha)\, e^{-\sqrt{12}\,\alpha} + F_2(\alpha) \mp F_2(\alpha)\}, \tag{44}
$$

where $\tau_0 = \frac{8}{5\sqrt{3}} \frac{m^2 R^3}{e^2 \gamma^5}$ is a characteristic time[4,5] associated with the spin flip radiation. For $R \approx 1$ km and $\gamma \approx 10^5$ then τ_0 will be of the order of one hour.

$$
F_1(\alpha) = \left(1 + \frac{41}{45}\alpha - \frac{23}{18}\alpha^2 - \frac{8}{15}\alpha^3 + \frac{14}{15}\alpha^4\right)
$$

$$
- \frac{8}{5\sqrt{3}} \left(1 + \frac{11}{12}\alpha - \frac{17}{12}\alpha^2 - \frac{13}{24}\alpha^3 + \alpha^4\right),
$$

$$
F_2(\alpha) = \frac{8}{5\sqrt{3}} \left(1 + \frac{14}{3}\alpha + 8\alpha^2 + \frac{23}{3}\alpha^3 + \frac{10}{3}\alpha^4 + \frac{2}{3}\alpha^5\right). \tag{45}
$$

One can see the exponential factor in (44), which appears for the same reasons as the one in (19). This is the Unruh-type contribution. Let us look at the phenomenon in greater detail. If $g = 2$ (i.e. $\alpha = 0$) we obtain the characteristic time equal to τ_0 and using $T_{\mp}$ from (44) and $F_{1,2}(\alpha)$ from (45) the equilibrium polarization takes the expected form[4, 5]

$$T_{\mp} \approx \frac{5\sqrt{3}}{8} \frac{e^2 \gamma^5}{m^2 R^3} \left(1 \mp \frac{8\sqrt{3}}{15} \right) \quad \text{and} \quad \mathcal{P}_0 = \frac{w_+ - w_-}{w_+ + w_-} = \frac{8}{5\sqrt{3}}. \quad (46)$$

If instead one considers the $g \gg 1$ limit then to lowest order, one obtains[4, 5]

$$T_+ \approx \frac{2}{3} \left| \frac{g}{2} \right|^5 \frac{1}{\tau_0}. \quad (47)$$

The first correction to (47) in this limit is due to the α^4/α^5 term in $F_1(\alpha)/F_2(\alpha)$. Using $T_{\mp}$ from (44) and $F_{1,2}(\alpha)$ from (45) gives to this next order the result

$$T_{\mp} \approx \frac{g^5}{2^6 \tau_0} \left\{ \left[\frac{14}{15} - \frac{8}{5\sqrt{3}} \right] \frac{e^{-\sqrt{3}\,(g-2)}}{g} + \frac{16}{15\sqrt{3}} \mp \frac{16}{15\sqrt{3}} \right\} + O\left(\frac{1}{g^2} \right). \quad (48)$$

In this limit, the equilibrium polarization is

$$\mathcal{P}_0 \approx \frac{1}{1 + \frac{F_1(\alpha)}{F_2(\alpha)} e^{-\sqrt{3}\,(g-2)}} \approx 1 - \left[\frac{\text{const.}}{g} + O\left(\frac{1}{g^2} \right) \right] e^{-\sqrt{3}\,g}. \quad (49)$$

Comparing this result with (20), we find perfect agreement if we take into account that here $\Delta\mathcal{E}/a \approx g/2$ and we are using the limit $g, \gamma \gg 1$.

4. The Equivalence Principle

In this section, we will examine the equivalence principle by comparing the detailed response of an Unruh–DeWitt detector undergoing uniform acceleration with an Unruh–DeWitt detector in a gravitational field. The theory of general relativity has as its conceptual underpinning the equivalence principle.[22] One version of the equivalence principle equates the inertial and gravitational mass — $m_{\text{inertial}} = m_{\text{grav}}$. This is called the weak equivalence principle and it leads to the result that the acceleration of an object in a gravitational field is independent of the nature of the object. This fact was already known to Galileo Galilei via several experiments, including the perhaps apocryphal story of the dropping of lead balls of different masses from the Tower of Pisa. A second version of the equivalence principle, and the one used by Einstein as the conceptual basis for general relativity, states that one cannot perform any local experiments which can

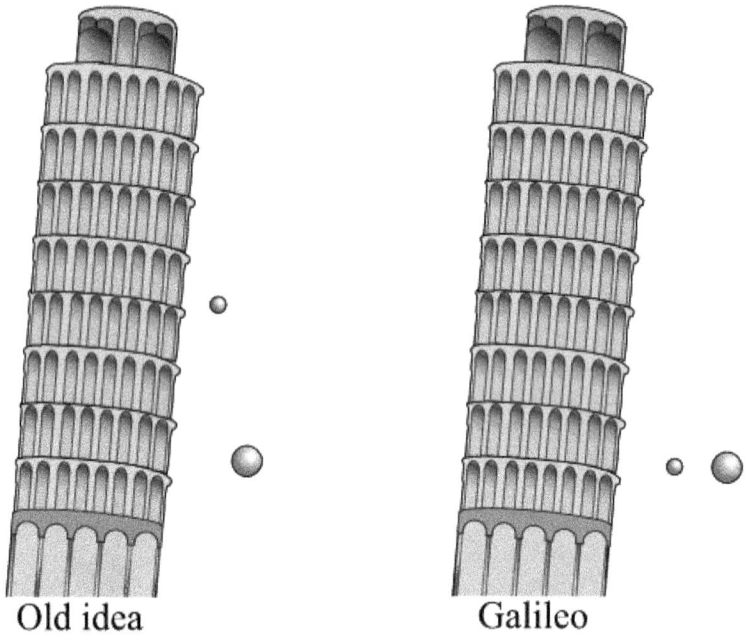

Fig. 4. The weak or Galiean equivalence Principle states that gravitational and inertial mass are equivalence and so all objects will fall with the same acceleration in a gravitational field regardless of composition.

distinguish between being in a gravitational field or being in an accelerating rocket or elevator. This second version of the equivalence principle is illustrated in Fig. 1 This restriction of locality is important since if one can perform experiments over extended distances or times then a gravitational field can be distinguished from an accelerating system via the tidal forces in the gravitational field. For example, two masses dropped from in the gravitational field of a spherical body like the Earth will move toward one another as they fall, whereas these same two masses dropped in an accelerating reference frame would fall parallel to one another. In this second version of the equivalence principle, the experiment or measurements must be in a region of space-time that is "small enough" which means in a region of space-time small enough to ignore the tidal effects of a real gravitational field.

In this chapter, we will probe the Equivalence Principle using the measurements of an Unruh–DeWitt detector placed in the space-time of a uniformly accelerating observer (i.e. Rindler space-time) and the space–time of a spherically symmetric, massive body (i.e. Schwarzschild space-time and compare the two measurements). We will check whether or not the Unruh–DeWitt detector measures radiation or not, and also we will investigate the details (response rate, temperature measured) of how the Unruh–Dewitt detector gets excited or not. We will find that whether radiation is detected or not (i.e. whether the Unruh-DeWitt detector gets excited or not) it will depend not only on the space-time and the space-time path through the space-time that the detector takes, but also depends on how one defines the vacuum state. Depending on which modes one uses to expand fields in terms of one can get different vacuum states and whether a particular observer detects particles or not depends crucially on which vacuum state they use. This will be explained in more detail in subsequent chapters. Some of the different vacua we will discuss are the Boulware vacuum, the Unruh vacuum, the Minkowski vacuum, the Hartle–Hawking vacuum and the Rindler vacuum[23-25] For example, we will find that for the Boulware vacuum that an Unruh–DeWitt detector will behave in the same way for an accelerating observer and for an observer in Schwarzschild space-time — in this case both detectors do not detect radiation/particles and thus the equivalence principle works in this case. However, next we compare an accelerating observer and an observe in a Schwarzschild space-time with respect to Minkowski vacuum, and while in this case both detectors detect radiation, there are differences in the details of this radiation which allows one to distinguish the accelerating observer from the observer in a gravitational field. This gives a violation of the equivalence principle.[26] We also find that in the limit when the detector in the Schwarzschild space-time approaches the horizon the violation of the equivalence principle goes away i.e. near the horizon the equivalence principle is restored.

One can ask what is the conceptual basis for this violation of the equivalence principle described above. The reason rests with the local nature of the equivalence principle versus the non-local nature of quantum phenomena. By making global space-time, measurements one can distinguish between a uniform acceleration and a gravitational field. For example, dropping two masses in an accelerating reference frame will lead to the masses falling parallel to one another so long as they are not stopped by the "floor" of the accelerating reference frame. Dropping the same two masses in a Schwarzschild space-time they will result in their moving towards one another because of the tidal forces associated with the spherically symmetric gravitating mass. In contrast to the local character of the equivalence principle, quantum mechanics has some inherent non-locality. The prime

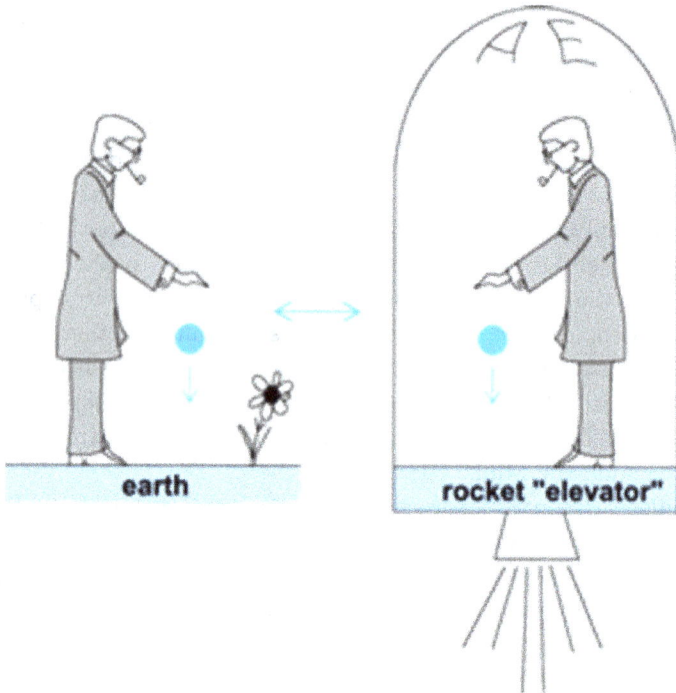

Fig. 5. The Einstein equivalence principle states that, locally, measurements/observations made in an accelerating reference frame and a gravitational field are equivalent.

example is Bell's inequality[27] experiments where particles have a non-local entanglement with one another. Also the definition of the vacuum mentions above is inherently non-local. To define a given vacuum state one expands a field in terms of the normal modes for a given space-time. For example, in flat, Minkowski space-time, these modes are non-local plane waves. Thus it is not surprising that quantum effects, like Hawking radiation or Unruh radiation, violate the equivalence principle. In some sense, the surprise is that the equivalence principle is restored as one approaches the event horizon, where the gravitation field is more intense as compared to when one if further from the horizon. There have been other suggestions of quantum phenomenon violating the equivalence principle. In Refs. 28–30, it is shown that neutrino oscillations violate the equivalence principle.

4.1. Minkowski space-time

As a warm up exercise to the two cases of interest we will study below (1+1-dimensional Rindler space-time of a uniformly accelerating observer and

1 + 1-dimensional Schwarzschild space-time for a stationary observer), we will first look at the response function for an inertial observer in Minkowski space-time.

In the case of motion with constant velocity v, the trajectory is $x(t) = (\gamma t, \gamma v t, 0, 0)$, with $\gamma = 1/\sqrt{1 - v^2}$ and t is the detector's proper time. Inserting this trajectory into (3), we obtain

$$\mathcal{F}_- \propto \lim_{\epsilon \to 0} \int_{-\infty}^{+\infty} d(\Delta\tau) \, \frac{e^{-i\Delta E \tau}}{(\gamma \, \Delta\tau - i\,\epsilon)^2 - (\gamma \, v \, \Delta\tau)^2}. \tag{50}$$

In (50), $\Delta E = E - E_0$. One can also write down $\mathcal{F}_+$, the transition rate per unit proper time for the detector decay from E to E_0. T_+ is given by

$$\mathcal{F}_+ \propto \lim_{\epsilon \to 0} \int_{-\infty}^{+\infty} d(\Delta\tau) \, \frac{e^{+i\Delta E \tau}}{(\gamma \, \Delta\tau - i\,\epsilon)^2 - (\gamma \, v \, \Delta\tau)^2}. \tag{51}$$

If instead of considering the phenomenon from the point of view of the co–moving reference frame we had considered the laboratory reference frame, $\mathcal{F}_\mp$ would be changed by a factor of γ because of time dilation. In (50) and (51), we are using the standard regularization via $i\epsilon$ which shifts the pole at $\Delta\tau = 0$ to the upper complex half-plane. As we will see in a moment such a regularization is necessary to have a non-zero probability for the spontaneous radiation of the detector if it is originally in the excited state.

The integral in (50) was already carried out in the case of 4D space-time in Sec. 2.1, with the full transition rate per unit time being given in (8). In terms of the response functions $\mathcal{F}_\pm$ one finds upon doing the contour integration with contours $C_\pm$ as defined in Sec. 2.1 the results

$$\mathcal{F}_- = 0, \quad \mathcal{F}_+ \propto \Delta\mathcal{E}. \tag{52}$$

$\mathcal{F}_-$ is zero because C_- does not enclose any poles. The contour C_+ does contain poles and thus $\mathcal{F}_+$ is non-zero. The results in (52) are independent of the velocity and, hence, are valid for a detector at rest $v = 0$ or moving with constant velocity. The physical meaning of the results for $\mathcal{F}_\pm$ is as follows: If the detector moves with constant velocity in the vacuum of a QFT there is zero probability for it to get excited, $\mathcal{F}_- = 0$. However, if the detector was originally in the excited state, there is a non-zero probability for it to radiate spontaneously, $\mathcal{F}_+ \neq 0$.

4.2. *Rindler space-time*

To simplify the calculations for the Rindler space-time (accelerating observer) and for Schwarzschild space-time (observer in a gravitational field) we examine 2D space-time. We do not lose any essential features of the response function in this way. As pointed out in Ref. 31 one of the key differences between the 2D response functions found in Refs. 3, 26 and 33 and the 4D response functions of Ref. 32, is that the prefactors of the 2D and 4D response functions are different. The temperatures obtained by the Unruh–DeWitt detectors from the 4D response function and the 2D response functions are exactly the same. It is because of this indepen-dence of temperature on the dimensionality of the space-time that we focus on temperature determined from the response function rather than on the whole response function. However as shown in Ref. 31 examining the full response function, $\mathcal{F}(E)$, still leads to a violation of the equivalence princi-ple. A more detailed discussion of the dependence of the response function on the dimensionality of the space-time is given in Ref. 34.

For both Rindler and Schwazschild space-times, we will use light front coordinates to write out the 2D space-time metrics in the form

$$ds^2 = C(u_i, v_i)du_i \, dv_i. \tag{53}$$

The index i indicates the particular space-time and vacuum being consid-ered, as described further below. The coordinates u_i and v_i will be taken to be the light front coordinates. For example, the light front coordinates for 2D Minkowski space-time are as follows:

$$u_{MM} = t - x, \quad v_{MM} = t + x,$$

with the conformal factor being one, $C(u, v) = 1$. The subscript $i = MM$ stands for Minkowski space-time and Minkowski vacuum. The Minkowski metric in these light front coordinates is

$$ds^2 = dt^2 - dx^2 = du_{MM}dv_{MM}.$$

For a detector at rest or moving with uniform velocity, v, one has $\Delta u_{MM} = \Delta v_{MM} = \Delta \tau$, with τ being proper time. If the detector is at rest $\Delta \tau = \Delta t$ while if the detector is moving $\Delta \tau = \Delta t \sqrt{1 - v^2}$.

Different forms of the light front metric lead to different forms of the wave equation which have different normal mode solutions — $e^{-i\omega u_i}$ and $e^{-i\omega v_i}$ where ω is the energy of the mode. Reference 3 gives the 2D Wight-man function for metrics of the form given in (53) as

$$G^+(x, t; x', t') = -\frac{1}{4\pi} \ln\left[(\Delta u_i - i\epsilon)(\Delta v_i - i\epsilon)\right]. \tag{54}$$

In (54), $\Delta u_i = u_i(x,t) - u_i(x',t')$ and $\Delta v_i = v_i(x,t) - v_i(x',t')$. For the 2D Minkowski space-time, Eq. (54) gives a Wightman function

$$G_{MM}^+ = -\frac{1}{4\pi} \ln\left[(\Delta\tau - i\epsilon)^2\right]. \tag{55}$$

(For a detector moving with uniform velocity v one must absorb a factor of $1/\sqrt{1-v^2}$ into ϵ.)

The issue, of which vacuum is being used to calculate the response function, is set by the form of the 2D Wightman function in (54), which in turn is set by the specific form of the metric. The recipe for determining the connection between the choice of metric and the specific vacuum is given as follows: (i) The form of the metric determines the specific form of the wave equation in the space-time which in turn determines the normal mode solutions, $u_k(x,t)$. (ii) A field, ϕ, can be expanded in terms of these modes as $\phi = \sum_k (a_k u_k + a_k^\dagger u_k^*)$. (iii) Turning the $a_k, a_k^\dagger$ into annihilation and creation operators then defines the vacuum e.g. $a_k|0_a\rangle = 0$; $a_k^\dagger|0_a\rangle = |1\rangle$. Now in flat, Minkowski space-time the construction of a vacuum is invariant under the action of the Poincaré group since Minkowski space-time has a symmetry associated with the Killing vector $\partial/\partial t$ which has eigenfunctions whose eigenvalues are $-i\omega$. For curved space-time or a space-time of a non-inertial observer there will in general not be a unique, observer independent way to construct normal modes or a vacuum state. In such a general space-time, there may be a second set of normal mode solutions, $v_k(x,t)$, and one may expand ϕ in terms of these new normal modes as

$$\phi = \sum_k (b_k v_k + b_k^\dagger v_k^*).$$

The new creation and annihilation operators, $b_k^\dagger, b_k$ define a new vacuum $b_k|0_b\rangle = 0$; $b_k^\dagger|0_b\rangle = |1\rangle$. Since both sets of normal modes are complete one can expand one set in terms of the other as

$$v_i = \sum_j (\alpha_{ij} u_j + \beta_{ij} u_j^*). \tag{56}$$

and

$$u_i = \sum_j (\alpha_{ji}^* v_j - \beta_{ji} v_j^*). \tag{57}$$

The relationships in (56) and (57) are the famous Bogoliubov transformations[35] and the coefficients α_{ij}, β_{ij} are the Bogoliubov coefficients which are given by

$$\alpha_{ij} = \langle v_i | u_j \rangle \quad \text{and} \quad \beta_{ij} = -\langle v_i | u_j^* \rangle. \tag{58}$$

Using all of the above, one can expand the annihilation operators for each set of normal modes as

$$a_i = \sum_j (\alpha_{ji} b_j + \beta_{ji}^* b_j^\dagger), \tag{59}$$

and

$$b_i = \sum_j (\alpha_{ij}^* a_j - \beta_{ij}^* a_j^\dagger). \tag{60}$$

Using (59) one can see that if $\beta_{ij} \neq 0$ then a_i will not be the annihilation operator for the vacuum state $|0_b\rangle$ i.e. $a_i |0_b\rangle \sum_j \beta_{ji}^* |1_b\rangle \neq 0$. Also the number operators $N_i = a_i^\dagger a_i$ is non-zero

$$\langle 0_b | N_i | 0_b \rangle = \sum_j |\beta_{ji}|^2 \neq 0. \tag{61}$$

Next we apply the above formalism to 2D Rindler space-time i.e. Minkowski space-time as seen by a uniformly accelerated observer with acceleration a. The path of such an observer is given by

$$t = \frac{1}{a} \sinh(a\tau); \quad x = \frac{1}{a} \cosh(a\tau), \tag{62}$$

where τ is the detector's proper time. The Rindler space-time in light front form is

$$ds^2 = dt^2 - dx^2 = du_{RM} dv_{RM}, \tag{63}$$

where

$$u_{RM} = t - x; \quad v_{RM} = t + x$$

giving

$$(\Delta u_{RM} - i\epsilon)(\Delta v_{RM} - i\epsilon) = (\Delta t - i\epsilon)^2 - (\Delta x)^2 .$$

The coordinates t and x are given by (62). The subscript RM stands for Rindler space-time and Minkowski vacuum state. This is explained in more detail below. Using (62) this becomes

$$\frac{1}{a^2}\left[(\sinh(a\tau) - \sinh(a\tau') - i\epsilon)^2 - ((\cosh(a\tau) - \cosh(a\tau'))^2\right]$$

$$= \frac{4}{a^2}\sinh^2\left(\frac{a(\Delta\tau - i\epsilon)}{2}\right) \tag{64}$$

Using the above results in (54) gives the Wightman function for this form of Rindler

$$G_{RM}^+ = -\frac{1}{4\pi}\ln\left[\frac{4}{a^2}\sinh^2\left(\frac{a(\Delta\tau - i\epsilon)}{2}\right)\right]. \tag{65}$$

Inserting this Wightman function into the response function (3), and performing a contour integration gives[3, 37] a Planckian response function

$$\mathcal{F}_{RM}(E) \propto \frac{1}{E(e^{E/k_BT_{RM}} - 1)} \quad \text{where } k_BT_{RM} = \frac{a}{2\pi}. \tag{66}$$

where k_B is Boltzmann's constant. This is the Unruh temperature given in terms of the acceleration, a, of the observer. The vacuum state associated with the form of the Rindler metric given by (63) is called the Minkowski vacuum 25, 36. It is with respect to this vacuum state that an observer will have a non-zero response function and will detect particles.

Rindler space-time can also be cast in Rindler coordinates (η, ξ) which are defined via

$$t = \frac{e^{a\xi}}{a}\sinh(a\eta); \quad x = \frac{e^{a\xi}}{a}\cosh(a\eta),$$

In these coordinates, the Rindler metric is

$$ds^2 = e^{2a\xi}(d\eta^2 - d\xi^2) = e^{2a\xi}du_{RR}dv_{RR}, \tag{67}$$

where the final form is in terms of light front coordinates

$$u_{RR} = \eta - \xi; \quad v_{RR} = \eta + \xi.$$

The subscript RR stands for Rindler space-time with respect the the Rindler vacuum state. The coordinate η plays the role of time and ζ plays the role of position. The proper time of the detector is $\tau = e^{a\xi}\eta$ and $ae^{-a\xi}$ is the proper acceleration. In this set of coordinates different ζ's correspond

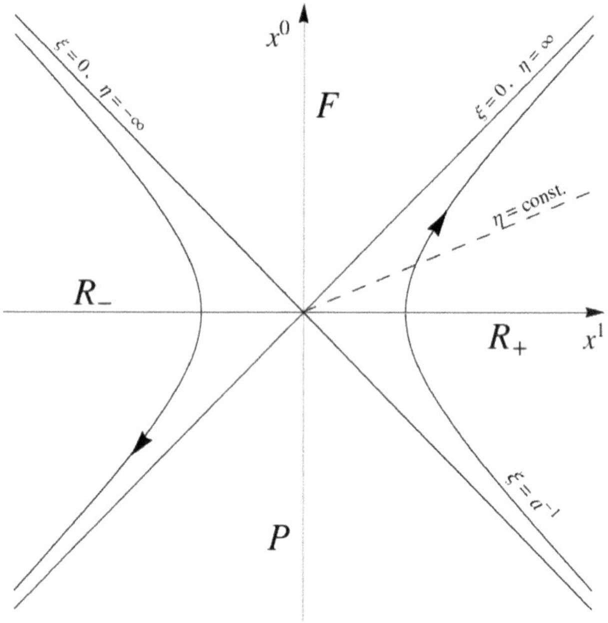

Fig. 6. Diagram of Rindler space-time. One has the future, F and past P wedges, as well as the left, R_-, and right wedges, R_+. The Rindler horizons are denoted by $\xi = 0$ and $\eta = \pm\infty$. The hyperbolic path of the accelerating observer is given by $\xi = a^{-1}$.

to picking different proper accelerations. Thus for a detector with a fixed proper acceleration $\xi = \text{const.}$ so that

$$\Delta u_{RR} = \Delta v_{RR} = \Delta \eta = e^{-a\xi}\Delta\tau.$$

The resulting Wightman function is then the same as that for 2D Minkowski space-time given in (55). One must now absorb a factor of $e^{a\xi}$ into the $i\epsilon$ term. Thus as for Minkowski space-time the response function is zero, $\mathcal{F}_{RR}(E) = 0$ — for comparison see $\mathcal{F}_-$ in Eq. (52). This Rindler vacuum state has been used[25] to preserve the equivalence principle against the following argument: "If a detector in uniform accelerated motion detects radiation but a detector at rest in a gravitational field without a horizon (e.g. a detector in the gravitational field of the Earth) does not detect radiation, can't one in this way distinguish a gravitational field from an acceleration?" The answer is that one needs to compare the two cases in question in the appropriate vacuum state — what Ginzburg and Frolov[25] call "corresponding vacua". Thus one should compare the Rindler vacuum of the accelerated observer with the Boulware vacuum (this vacuum state is discussed next) of the detector fixed in a gravitational field without a horizon. In this way the equivalence principle is preserved against the preceding argument.

4.3. *Schwarzschild space-time*

We now calculate the response function for detector in 2D Schwarzschild space-time. Starting from the standard form of the metric

$$ds^2 = (1 - 2M/r)dt^2 - (1 - 2M/r)^{-1}dr^2$$

(where we have taken $G = 1$ and $c = 1$) we transform this into the light front form

$$ds^2 = \left(1 - \frac{2M}{r}\right) du_{SB} \, dv_{SB}. \tag{68}$$

In this equation

$$u_{SB} = t - r^*; \quad v_{SB} = t + r^*$$

with

$$r^* = \int \frac{dr}{1 - 2M/r} = r + 2M \ln |r/2M - 1|.$$

The subscript SB stands for Schwarzschild space-time in the Boulware vacuum state. For a detector at a fixed radius $r = R$, $r^* \to R^* = R + 2M \ln(R - 2M)$ and so $\Delta r^* = 0$. Thus

$$\Delta u_{SB} = \Delta v_{SB} = \Delta t.$$

The relationship between proper time, τ, and Schwarzschild time, t, is

$$\Delta \tau = \sqrt{1 - \frac{2M}{R}} \Delta t.$$

Combining these results we see that the Wightman function for the Schwarzschild metric of the form (68) is essentially the same as the Minkowski space-time Wightman function in (55). One must absorb the constant factor $\sqrt{1 - 2M/R}$ into ϵ. Thus as for Minkowski space-time the response function for this form of the Schwarzschild metric is zero, $\mathcal{F}_{SB}(E) = 0$ — compare to $\mathcal{F}_-$ from Eq. (52). This result may seem surprising since it appears we have shown Hawking radiation does not exist. Actually what this shows is that there is no radiation with respect to the vacuum state defined by the choice of the Schwarzschild metric in (68). This vacuum state (68) is called the Boulware vacuum.[23] There is no Hawking radiation with respect to the Boulware vacuum. However, the Boulware vacuum is not physical near the horizon since in the Boulware vacuum the energy–momentum tensor diverges at the horizon.

Two vacuum states which are well behaved at the horizon are the Hartle–Hawking vacuum[24] and Unruh vacuum.[1] To study these two vacuum states, we write the Schwarzschild metric in Kruskal form

$$ds^2 = \frac{2M}{r} e^{-r/2M} du_{SH} dv_{SH},$$

with

$$u_{SH} = -4M e^{-u_{SB}/4M} \quad ; \quad v_{SH} = 4M e^{v_{SB}/4M}$$

where u_{SB}, v_{SB} were previously defined and $r = r(u_{SH}, v_{SH})$ is an implicitly defined function of u_{SH}, v_{SH}. The subscript SH stands for the Schwarzschild space-time in the Hartle–Hawking vacuum. The Hartle–Hawking vacuum uses the coordinates, u_{SH}, v_{SH}, to calculated $\Delta u_{SH}, \Delta v_{SH}$ which are then used to calculate the associated Wightman function. The Hartle–Hawking vacuum corresponds to the state when a Schwarzschild black hole is in equilibrium with a thermal bath which is at the same temperature as that of the black hole. For a realistic black

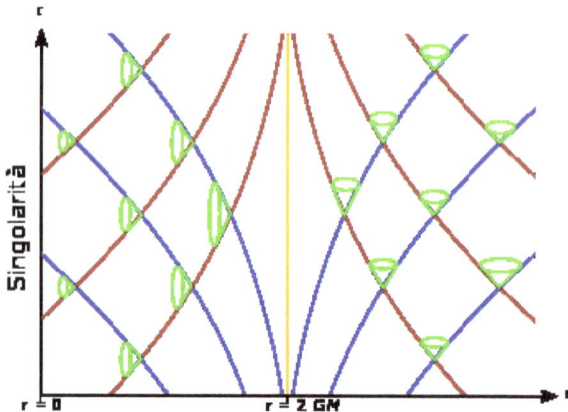

Fig. 7. The 2D Schwarzschild space-time with horizon at $r = 2GM$. The red and blue lines are incoming and outgoing null geodesics and the green light cones are shown at different space-time points. At the horizon the light cones "open" into the horizontal direction.

hole formed via collapse the Unruh vaccum is a better choice. The Unruh vacuum corresponds to the Boulware vacuum in the far past and the Hartle–Hawking vacuum in the far future. In detail for the Unruh vacuum a detector fixed $r = R$ has $\Delta u_{SU}, \Delta v_{SU}$ given by[3]

$$\Delta u_{SU} = \Delta u_{SH} = -4Me^{R^*/4M}(e^{-t/4M} - e^{-t'/4M}),$$

$$\Delta v_{SU} = \Delta v_{SB} = \Delta t. \tag{69}$$

R^* is defined by setting $r = R$ in the expression for r^* and $\Delta t = t - t'$. Using (69) gives

$$(\Delta u_{SU} - i\epsilon)(\Delta v_{SU} - i\epsilon)$$

$$= 8Me^{R^*/4M}e^{-(t+t')/8M}(\Delta t - i\epsilon)\sinh\left(\frac{\Delta t - i\epsilon}{8M}\right)$$

$$= 8Me^{R^*/4M}e^{-(t+t')/8M}\left(\frac{\Delta\tau - i\epsilon}{\sqrt{1-2M/R}}\right)\sinh\left(\frac{\Delta\tau - i\epsilon}{8M\sqrt{1-2M/R}}\right). \tag{70}$$

Inserting this result into (54) gives the Wightman function which is essentially the same as the Wightman function for the Rindler metric in the Minkowski vacuum (55) but with a replaced by $1/(4M\sqrt{1-2M/R})$, which is the blue shifted surface gravity $\kappa = 1/4M$ of the black hole. Inserting this Wightman function in the response function (3) then leads to the same type of contour integral as before. The factor $e^{-(t+t')/8M}$ does not contribute to the contour integration and the multiplicative factor, $\Delta\tau - i\epsilon$, is essentially that for Minkowski space-time and also does not contribute. The result is again a Planckian spectrum

$$\mathcal{F}_{SU}(E) \propto \frac{1}{E(e^{E/k_B T_{SU}} - 1)} \quad \text{with} \quad k_B T_{SU} = \frac{1}{8\pi M\sqrt{1-2M/R}} \tag{71}$$

which is the temperature measured by the fixed detector with respect to the Unruh vacuum for Schwarzschild space-time. Doing the calculation in the Hartle–Hawking vacuum yields the same temperature. The only difference is that Eq. (70) for the Hartle–Hawking vacuum is proportional to $\sinh^2(\dots)$ rather than $\sinh(\dots)$. This leads to an unimportant, multiplicative factor of two in $\mathcal{F}_{SH}(E)$. This result for T_{SU} from (71) is consistent with the higher dimensional embedding approach of.[38,39] In these works, the temperature of the static Schwarzschild observer in (71) is obtained as the Unruh temperature in a six-dimensional space-time using an effective six-acceleration equals the blue shifted surface gravity, $\kappa = 1/(4M\sqrt{1 - 2M/R})$. The magnitude of the real acceleration measured by this fixed observer, $r = R$, in the Schwarzschild space-time, is given by[40]

$$a_S = \frac{\sqrt{\nabla_\mu V \nabla^\mu V}}{V} = \frac{M}{R^2\sqrt{1 - 2M/R}}, \tag{72}$$

where $V = \sqrt{1 - 2M/r}$ is the redshift factor for the Schwarzschild space-time. A similar calculation for the Rindler space-time (using the Schwarzschild-like form of the Rindler metric $ds^2 = (1 + 2ax)dt^2 - (1 + 2ax)^{-1}dx^2$) yields $a_R = a$ as expected. It is the results in (71) and (72) which lead to a violation of the equivalence principle when compared to similar measurements in Rindler space. Both the fixed observer in Schwarzschild and the observer in Rindler space should measure the same acceleration so we set $a_R = a = a_S$ where a is the acceleration of the Rindler observer. If we then substitute this into (66) we get

$$k_B T_{RM} = \frac{M}{2\pi R^2 \sqrt{1 - 2M/R}}. \tag{73}$$

By comparing (71) and (73) it is clear that $T_{RM} < T_{SU}$ when $R > 2M$. Thus one can tell the two systems apart by making local space-time measurements. As one approaches the horizon, $R = 2M$, the two results converge to the same (infinite) value. Thus at the horizon the equivalence principle is restored. The procedure for an observer equipped with a means of measuring local acceleration and an Unruh–Dewitt detector for measuring temperature is as follows: (i) Measure the local acceleration. (ii) Insert this into (66). (iii) Measure the temperature via the Unruh–Dewitt detector. If the temperature is higher than that calculated in step (ii) then one is in a gravitational field and not in an accelerating frame.

4.4. Mach's principle, equivalence principle, and gravity

The fact that inertial mass and gravitational mass are the same, within experimental limits, is one of the "coincidences" in physics. This fact underlies the weak equivalence principle or Galilean equivalence principle. It leads to the result that all objects, regardless of composition will fall with the same acceleration in a gravitational field. Briefly one as

$$m_{\text{inert}} = m_{\text{grav}}$$

Now from Newton's second law of motion one finds that the force and acceleration of an object are related via $F = m_{\text{inert}} a$. Now for an object in a gravitational field g, Newton's law of gravitation gives the gravitational force on the object as $\mathbf{F} = m_{\text{grav}} \mathbf{g}$. Equating these two forces *and* using the equivalence $m_{\text{inert}} = m_{\text{grav}}$ one finds that $\mathbf{a} = \mathbf{g}$ i.e. all objects experience the same acceleration.

In constructing his theory of gravity Einstein elevated this equivalence to the central functional principle of general relativity.[41] The fact of being based on an underlying physical principle distinguishes general relativity from quantum mechanics. Quantum mechanics does not have a similar physical principle at its foundations. One potential principle that *might* play the same role for quantum mechanics that the equivalence principle plays for general relativity is the recent and loosely formulated "holographic principle".[42] One hope is that the holographic principle might provide a path to construct a consistent theory of quantum gravity.

A second formulation of the equivalence principle — and the one used chiefly in this work — is called the Einstein equivalence principle and makes use of the Einstein elevator.[41] If an observer placed inside a small enough enclosure, such as an elevator, feels their feet pressed to the bottom of the elevator, this observer cannot tell if the elevator is at rest on the surface of some gravitating planet, with a local gravitation field $\mathbf{g}$, or if the elevator is accelerating through empty space-time with acceleration $\mathbf{a} = \mathbf{g}$. The condition "small enough" means that tidal forces, which *do* differentiate between a gravitational field and acceleration, can be ignored. This locality condition gives general relativity a local character.

In counterpoint Mach's principle, which is said to have inspired Einstein, does not share the same central role as the equivalence principle in the foundations of general relativity. This is partly due to the fact that Mach's principle is vaguely defined. In this essay we will take Mach's principle to have the following meaning: "The inertial properties of an object are determined by the energy–momentum throughout all space".[43] This statement makes it clear that Mach's principle is, at least partly, a global principle — the local inertial properties of a particle are determined by the global distribution of energy–momentum throughout the entire space. Given this tension between the global nature of Mach's principle versus the local character of the equivalence principle, it might not be surprising that only one of these principles has found a firm footing in the foundations of general relativity. However, by comparing the temperatures of Hawking and Unruh radiation we will find support for a *thermal* Mach's principle and, as shown above there is a violation of the Einstein elevator version of the equivalence principle.

4.5. *Thermal Mach's principle*

One way to understand the reason for the violation of the Einstein elevator version of the equivalence principle is that the equivalence principle is fundamentally a local principle — a gravitational field and acceleration are only equivalent in a small enough space-time region. In contrast Hawking and Unruh radiation depend on how one defines the quantum vacuum[3] which is a non-local construction. Briefly, to describe this construction, suppose one has a scalar field, ϕ, which obeys the wave equation $\Box_g \phi = 0$ where the subscript g indicates the type of space-time one has (e.g. Minkowski, Rindler, Schwarzschild, *etc.*). The solution to $\Box_g \phi = 0$ has a set of mode solutions $u_k(x)$ where the subscripts k are the momenta of the particular field mode. One can then expand a general field ϕ in terms of these modes as $\phi(x) = \sum_k [a_k u_k(x) + a_k^\dagger u_k^*(x)]$ where $a_k, a_k^\dagger$ are annihilation/creation operators of quanta of momenta k. The operators $a_k, a_k^\dagger$ define the vacuum state. Note also that the modes, $u_k(x)$, are non-local in extent. In any case whether one has Minkowski vacuum, Rindler vacuum, Boulware vacuum, Hartle–Hawking vacuum, the vacuum state is a non-local construction. Since Hawking and Unruh radiation depend on non-local vacuum states, it is not surprising that one finds these effects violate the equivalence principle which is local.

The discussion above points to a *thermal* Mach's principle i.e. the system's local thermal properties such as the temperature, radiation spectrum, and entropy depend on the non-local nature of the quantum vacuum of the space-time. This has similarity to the statement quoted earlier from Ref. 43 where the local inertial properties of the system are determined by the global distribution of the energy–momentum in the space-time.

One could then argue that the above-proposed thermal Mach's principle sheds light on the original Mach's principle where the local inertial properties are determined by the global energy–mass distribution. The agreement of the temperatures, $T_{\text{Hawking}} = T_{\text{Unruh}}$, in the limit as one approaches the horizon, $r \to 2GM$, fits nicely with the proposed holographic models of Refs. 48 and 49 where inertia arises from the entropy of the system. Combining the thermal Mach's principle with the idea from these holographic models, that inertia comes from thermodynamic quantities such as entropy, may suggest the original, inertial version of Mach's principle as a conceptual basis for a quantum theory of gravity.

5. Summary and Conclusions

In this chapter, we have studied how an Unruh–DeWitt detector can be used to study particle creation in a given space-time. The Unruh–DeWitt detector is a very concrete, physical, and simple way to probe this issue. In its simplest form, Unruh–DeWitt detector is a quantum, two-state system which is coupled to some quantum field (in this work, we have taken the case of an Unruh–DeWitt detector coupled to a scalar field). In the first part of this work, we showed how one could experimentally confirm the Unruh effect using electrons in storage rings. Electrons in storage rings such as DESY (see Fig. 2) undergo large centripetal accelerations on the order of 10^{23}–$10^{24}\frac{m}{s^2}$. From (11) accelerations of this magnitude would imply a Unruh temperature in 10^3–10^4 kelvin. Such a temperature would have noticeable consequences. Specifically, the electrons in storage rings are in a magnetic field and thus have two energy states, making them natural Unruh–DeWitt. Due to the being exposed to an effective temperature of a few 10^3 kelvin, some fraction of the electrons will be excited to the higher energy state (i.e. the state with the electron's spin aligned with the magnetic field). A subtle point in this analysis is the electrons are undergoing circular acceleration, rather than linear acceleration, the spectrum of particles the electrons encounter is not a thermal/Planckian spectrum. From the lab frame, the electrons undergo transition to the upper energy state, due to spin flip radiation. Viewed in this way the occupation of some fraction of the electrons in the upper energy state is known as the Sokolov–Ternov Effect.[13] In Sec. 2 and 3, we showed that the equilibrium polarization of the depolarization of electrons seen as an Unruh effect in the rest frame of the electrons (Eq. (20)) versus the depolarization of the electrons seen via the spin flip radiation as in the Sokolov–Ternov effect (Eq. (49)) had exactly the same form. This then provides experimental evidence for the Unruh effect.

In the second half of this chapter, we used the Unruh–DeWitt detector to show that there is a violation of the Einstein elevator version of the equivalence principle i.e. when one compares the temperature of a detector fixed at $r = R > 2M$ in a Schwarzschild black hole space-time and a uniformly accelerating detector, for the same locally measured acceleration. Both detectors will detect thermal radiation, but for equal, local accelerations the detector in the gravitational background will measure a higher temperature than the accelerating detector. The subtle feature in the analysis arises because one must compare the response function of the detector with respect to what Ginzburg and Frolov[25] call "corresponding" or "matched" vacua. When comparing an accelerating detector in the Rindler vacuum with a fixed detector in Schwarzschild space-time in the Boulware vacuum, both detectors will not detect radiation. On the other hand, comparing an accelerating detector in Minkowski vacuum with a fixed detector in Schwarzschild space-time in Unruh vacuum or Hartle-Hawking vacuum, both detectors will detect thermal radiation. Up to this point, the equivalence principle works qualitatively. However, if one compares the values of the temperature of the thermal radiation measured in each case — Eq. (66) for the accelerating detector and Eq. (71) for the detector fixed in a gravitational field — one finds that for the same acceleration the detector in gravitational field will measure a higher temperature. This allows one to tell the two cases apart. As $R \to \infty$, $T_{SU} \to 1/8\pi M$ while from (73) $T_{RM} \to 0$. This latter result occurs since as $R \to \infty$ the acceleration due to gravity goes to 0 as $1/R^2$ i.e. neither the acceleration from (72) nor the associated temperature T_{RM} from (73) are long range. Hawking radiation is long range since it falls off like $1/R$, and thus does have a constant flux/temperature as $R \to \infty$. Conversely as one approaches the horizon the two temperatures, T_{SU} and T_{RM}, approach the same (infinite) value. One can interpret this as quantum field theory/quantum mechanics becoming more compatible with the equivalence principle and general relativity as one approaches the extreme conditions near a black hole horizon. This could optimistically be taken as a hint that gravity and quantum mechanics become more compatible and begin to merge into a consistent theory of quantum gravity at high energies/extreme gravitational fields.

Acknowledgments

The work of ETA was supported by the grant from the Foundation for the Advancement of Theoretical Physics and Mathematics "BASIS, by RFBR grants 19-02-00815 and 21-52-52004, and by Russian Ministry of Education and Science.

References

1. W. G. Unruh, *Phys. Rev. D* **14**, 870 (1976).
2. B. S. DeWitt, in eds. S. W. Hawking and W. Israel, *General Relativity: An Einstein Centenary Survey*, pp. 680–745. Cambridge University Press, Cambridge, (1979).
3. N. D. Birrell and P. C. W. Davies, *Quantum Fields in Curved Space*, Cambridge University Press, Cambridge (1982).
4. E. T. Akhmedov and D. Singleton, *Int. J. Mod. Phys. A* **22**, 4797 (2007).

5. E. T. Akhmedov and D. Singleton, *JETP Lett.* **86**, 615 (2007).
6. L. C. B. Crispino, A. Higuchi and G. E. A. Matsas, *Rev. Mod. Phys.* **80**, 787 (2008).
7. J. Louko and A. Satz, *Class. Quantum Grav.* **23**, 6321 (2006).
8. J. Louko and A. Satz, *Class. Quantum Grav.* **25**, 055012 (2008).
9. S. Schlicht, *Class. Quantum Grav.* **21**, 4647 (2004).
10. J. I. Korsbakken and J. M. Leinaas, *Phys. Rev.* D **70**, 084016 (2004).
11. I. M. Frank, *Izv. Akad. Nauk SSSR, Ser. Fiz.* **6**, 3 (1942); V. P. Frolov and V. L. Ginzburg, *Phys. Lett.* A **116**, 423 (1986).
12. L. Sriramkumar and T. Padmanabhan, *Int. J. Mod. Phys.* D **11**, 1 (2002); arXiv:gr-qc/9903054; K. Srinivasan, L. Sriramkumar and T. Padmanabhan, *Phys. Rev.* D **58**, 044009 (1998); arXiv:gr-qc/9710104; T. Padmanabhan, *Astrophys. Space Sci.* **83**, 247 (1982).
13. A. A. Sokolov and I. M. Ternov, *Dok. Akad. Nauk. SSSR.* **153**, 1052 (1963); (in Russian) A. A. Sokolov and I. M. Ternov, *Sov. Phys. Dokl.* **8**, 1203 (1964).
14. J. R. Letaw and J. D. Pfautsch, *Phys. Rev.* D **22**, 1345 (1980); J. R. Letaw, *Phys. Rev.* D **23**, 1709 (1981).
15. J. S. Bell and J. M. Leinaas, *Nucl. Phys.* B **212**, 131 (1983); J. S. Bell and J. M. Leinaas, *Nucl. Phys.* B **284**, 488 (1987).
16. D. Müller, arXiv:gr/qc/951203.
17. G. Volovik, *Int. Ser. Monogr. Phys.* **117**, (2006).
18. J. D. Jackson, *Rev. Mod. Phys.* **48**, 417 (1976).
19. L. Landau and L. Lifshits, *Relativistic Quantum Theory*, Vol. IV, Elsevier Science Ltd. (1977).
20. A. Borisov, A. Sokolov, I. Ternov and V. Zhukovski, *Quantum Electrodynamics*, Izdatelstvo Moskovskogo Universiteta (1983) (in Russian).
21. L.Landau and L.Lifshits, *The Classical Theory of Fields*, Vol. II, Elsevier Science Ltd. (1977).
22. A. Einstein, *Ann. Phys.* **11**, 898 (1911).
23. D.G. Boulware, *Phys. Rev.* D **11**, 1404 (1975); *Phys. Rev.* D **12**, 350 (1975).
24. J. B. Hartle and S.W. Hawking, *Phys. Rev.* D **13**, 2188 (1976).
25. V. L. Ginzburg and V. P. Frolov, *Sov. Phys. Usp.* **30**, 1073 (1987).
26. D. Singleton and S. Wilburn, *Phys. Rev. Lett.* **107**, 081102 (2011).
27. J. S. Bell, *Physics* **1**, 195 (1964).
28. M. Gasperini, *Phys. Rev.* D **38**, 2635 (1988); *Phys. Rev.* D **39**, 3606 (1989).
29. G. Z. Adunas, E. Rodriguez-Milla and D. V. Ahluwalia, *Phys. Lett.* B **485**, 215 (2000); *Gen. Rel. Grav.* **33**, 183 (2001).
30. J. R. Mureika, *Phys. Rev.* D **56**, 2408 (1997).
31. L. C. B. Crispino, A. Higuchi and G. E. A. Matsas, *Phys. Rev. Lett.* **108**, 049001 (2012).
32. P. Candelas, *Phys. Rev.* D **21**, 2185 (1980).
33. D. Singleton and S. Wilburn, *Phys. Rev. Lett.* **108**, 049002 (2012).
34. L. C. B. Crispino, A. Higuchi and G. E. A. Matsas, *Phys. Rev.* D **70**, 127504 (2004).
35. N. N. Boboliubov, *Sov. Phys. JETP* **7**, 51 (1958) [*Zh. Eksp. Teor. Fiz.* **34**, 58 (1958)].

36. L. P. Grishchuk, Ya. B. Zel'dovich and L. V. Rozhanskii, *Sov. Phys. JETP* **65**, 11 (1987).
37. R. Brout *et al.*, *Phys. Rep.* **260**, 329 (1995).
38. S. Deser and O. Levin, *Class. Quantum Grav.* **15**, L85 (1998); *Phys. Rev. D* **59**, 064004 (1999).
39. E. J. Brynjolfsson and L. Thorlacius, *JHEP* **09**, 066 (2008).
40. S.M. Carroll, *An Introduction to General Relativity Spacetime and Geometry*, Section 6.3, Addison-Wesley, San Francisco (2004).
41. A. Einstein, *Ann. Phys.* **11**, 898 (1911).
42. G. Hooft, arXiv:gr-qc/9310026; L. Susskind, *J. Math. Phys.* **36**, 6377 (1995).
43. J. A. Wheeler, in eds. H.-Y. Chiu and W. F. Hoffman, *Gravitation and Relativity*, p. 303. Benjamin, New York (1964).
44. S.W. Hawking, *Comm. Math. Phys.* **43**, 199 (1975).
45. L. C. Barbado, C. Barcelo and L. J. Garay, *Class. Quantum Grav.* **28**, 125021 (2011); L. C. Barbado, C. Barcelo and L. J. Garay, *Class. Quantum Grav.* **29**, 075013 (2012).
46. S.D. Bièvre and M. Merkli, *Class. Quantum Grav.* **23**, 6525 (2006).
47. A. Almheiri, D. Marolf, J. Polchinski and J. Sully, *JHEP* **1302**, 062 (2013).
48. E. Verlinde, *JHEP* **1104**, 029 (2011).
49. M. Gogberashvili and I. Kanatchikov, *Int. J. Theor. Phys.* **51**, 985 (2012).
50. K.K. Ng, R. B. Mann and E. Martin-Martinez, arXiv: 1606.06292 [quant-ph]

Chapter 3

Gravity at Finite Temperature, Equivalence Principle, and Local Lorentz Invariance

Maurizio Gasperini

Dipartimento di Fisica, Università di Bari,
Via G. Amendola 173, 70126 Bari, Italy
Istituto Nazionale di Fisica Nucleare, Sezione di Bari, Italy
maurizio.gasperini@ba.infn.it

In this chapter, we illustrate the close connection between the violation of the weak equivalence principle typical of gravitational interactions at finite temperature, and similar violations induced by a breaking of the local Lorentz symmetry. We also discuss the physical implications of the effective repulsive forces possibly arising in such a generalized gravitational context, by considering, for an illustrative purpose, a quasi-Riemannian model of gravity with rotational symmetry as the local gauge group in tangent space.

1. Introduction

A breakdown of the Einstein's equivalence principle, which is the main subject of this book, is expected to occur also in the context of gravity at finite temperature, at the level of both classical/macroscopic and microscopic/quantum field interactions: in both cases, there are indeed deviations from the standard, geodesic time-evolution and the locally-inertial type of motion. There are, however, important differences between the two cases.

In the case of classical test bodies, one finds thermal geodesic deviations which depend on the mass and total energy of the body,[1] and which can be described by an effective geometry of the standard "metric" type but with broken local Lorentz symmetry. The geodesic deviations at the quantum/microscopic level, on the contrary, are found to correspond to an effective *locally hyperbolic* type of free motion[2] (characterized by a constant, nonvanishing, tangent-space acceleration) and require, for their classical description, a Lorentz invariant but "metric-affine" (or Weyl) generalized geometrical structure.

In this chapter, we will concentrate on the first type of temperature-dependent effects, and we will show that the corresponding non-geodesic motion of massive, point-like bodies is just a particular case of more general equations of motion predicted by matter–gravity interactions which are not locally Lorentz invariant, but only locally SO(3)-invariant.[3-5] This suggests that an efficient classical description of macroscopic gravity at finite temperature may be successfully implemented in the context of an effective geometric model different from General Relativity, and in which the local gauge symmetry of the SO(3, 1) group is broken also in the action describing the free gravitational dynamics.

A possible example of such a gravitational theory is provided by the so-called *quasi-Riemannian* models of gravity,[6-9] and in particular by that class of models with local rotational symmetry in tangent space.[3, 10, 11] Such an unconventional geometric structure is motivated by the fact that, at finite temperature, the flat tangent manifold describing the Minkowski vacuum has to be replaced by a tangent *thermal bath* at finite, non-vanishing temperature,[2] which breaks the general SO(3, 1) invariance but preserves the local SO(3) symmetry in the preferred rest frame of the heat bath.

A gravitational model of this type may have interesting cosmological applications. In fact, the violation of the equivalence principle due to the breaking of the local Lorentz symmetry may be associated with the presence of effective repulsive interactions,[4, 9, 12, 13] whose consequences are relevant for both inflationary models[11] and bouncing models preventing the initial singularity.[3, 12] This anticipates, in particular, scenarios very similar to the ones arising in a string cosmology context (see e.g. Refs. 14–17).

In this chapter, we will present a short review of the above results obtained in previous papers and illustrating the close connection between gravity at finite temperature, gravitational interactions with broken local Lorentz symmetry, and violation of the weak equivalence principle. We will take the opportunity of clarifying some technical details, not explicitly mentioned in the previous literature. The chapter is organized as follows.

We will start in Sec. 2 with the mass-dependent deviations from geodesic motion for free-falling test particles at finite temperature, and we shall discuss, in Sec. 3, the more general form of such deviations for gravitational interactions with broken local Lorentz symmetry. A simple, general covariant but locally only SO(3)-invariant model of gravity, and its possible cosmological consequences, will be briefly discussed in Sec. 4. A few concluding remarks will be finally presented in Sec. 5.

2. Non-geodesic Motion of Test Particles at Finite Temperature

Let us start by recalling that at finite temperature inertial and gravitational masses are in general different,[18,19] as they correspond, from a thermodynamical point of view, to the low momentum limit of free energy and internal energy, respectively.[20]

Considering in particular a charged particle of proper mass m_0, in thermal equilibrium with a photon heat bath at a temperature $T \ll m_0$, and in the absence of gravity, one finds that its total free energy E can be written (to lowest order in T^2) as[18,19]

$$E(T) = \left[m_0^2 + p^2 + \frac{2}{3} \alpha \pi T^2 \right]^{1/2} \equiv \left[m^2(T) + p^2 \right]^{1/2},$$

$$m(T) = \left[m_0^2 + \frac{2}{3} \alpha \pi T^2 \right]^{1/2} \simeq m_0 \left(1 + \frac{1}{3} \alpha T^2 \right), \tag{1}$$

where α is the fine structure constant and m_0 the (renormalized) rest mass at $T = 0$. We are using units in which $\hbar$, c and the Boltzmann constant k_B are set equal to one.

On the other hand, according to the results of a detailed finite-temperature calculation performed in the weak field limit and in the rest frame of the heat bath,[18,19] it turns out that the effective energy–momentum $\theta^{\mu\nu}$, representing the contribution of the same particle to the "right-hand side" of the gravitational Einstein equations, can be expressed (again, to lowest order in T^2) as follows[1]:

$$\theta^{\mu\nu} = T^{\mu\nu} - \frac{2}{3} \alpha \pi \frac{T^2}{E^2(T)} V_4^\mu V_4^\nu T^{44}. \tag{2}$$

Here $T^{\mu\nu}$ is the standard, minimally coupled to gravity, particle stress tensor (with temperature-dependent mass term), T^{44} is the corresponding energy–density component locally evaluated in the flat tangent-space limit, and $E(T)$ is the (temperature-dependent) energy of Eq. (1). Finally, V_a^μ is the so-called vierbein (or tetrad) field, connecting the world metric $g^{\mu\nu}$ of the given Riemann manifold to the flat Minkowski metric η^{ab} of the local tangent space, in such a way that $g^{\mu\nu} = V_a^\mu V_b^\nu \eta^{ab}$. It follows that

$$T^{44} = T^{\alpha\beta} V_\alpha^4 V_\beta^4, \tag{3}$$

and that the energy E of Eq. (1), representing the time-like component of the particle four-momentum at finite temperature in the locally flat space-time limit, can be related to the components of the "curved" (i.e. generally covariant) momentum p^μ by

$$E \equiv m\dot{x}^4 \equiv p^4 = p^\mu V_\mu^4 = m\dot{x}^\mu V_\mu^4 \tag{4}$$

(a dot denotes differentiation with respect to the proper time τ).

It is appropriate, at this point, to clearly specify the adopted conventions. We will use Latin letters to denote flat (also called *anholonomic*) tangent space indices, and Greek letters to denote general-covariant (*holonomic*) world indices. Also, the index "4" will always refer to the time-like coordinate x^4 of tangent space, while the index "0" to the time-like coordinate $x^0 \equiv t$ of the curved Riemann manifold.

Given the above definitions and conventions, it is clear that the energy–momentum (2) correctly transforms according to the tensor representation of the general diffeomorphism group $x^\mu \to x'^\mu(x)$ acting on the coordinates of the Riemann manifold, but is not a scalar object under the action of the Lorentz symmetry group in the local Minkowski tangent space. However, the energy–momentum (2) is manifestly compatible with a local rotational symmetry: the invariance under local transformations of the O(3) group acting on the flat (Latin) indices. This is consistent with the presence in the local tangent space of a preferred frame at rest with the heat bath, which is indeed the frame where the explicit form of the effective gravitational source (2) has been computed, and where the thermal radiation is isotropically distributed.

Let us also assume, for the moment, that any *direct* modification of the free geometric dynamics due to the temperature is absent (or negligible), so that the "left-hand side" of the Einstein equations keeps unchanged, and the effective gravitational equations at finite temperature take the form $G^{\mu\nu} = 8\pi G\, \theta^{\mu\nu}$, where $G^{\mu\nu}$ is the usual Einstein tensor and G the Newton constant. The contracted Bianchi identity $\nabla_\nu G^{\mu\nu} = 0$, where ∇_ν is the Riemann covariant derivative, thus implies the generalized conservation equation $\nabla_\nu \theta^{\mu\nu} = 0$, which for the effective energy–momentum tensor (2) can be written explicitly as follows:

$$\partial_\nu(\sqrt{-g}\, T^{\mu\nu}) - \frac{2}{3}\alpha\pi\, \partial_\nu\left(\sqrt{-g}\, \frac{T^2}{E^2} V_4^\mu V_4^\nu T^{\alpha\beta} V_\alpha^4 V_\beta^4\right)$$

$$+ \sqrt{-g}\, \Gamma_{\nu\alpha}{}^\mu\left(T^{\alpha\nu} - \frac{2}{3}\alpha\pi\, \frac{T^2}{E^2} V_4^\alpha V_4^\nu T^{\beta\rho} V_\beta^4 V_\rho^4\right) = 0. \tag{5}$$

We should now recall that, given a test body and the covariant conservation of its energy–momentum tensor, the corresponding equation of motion can be obtained by applying the so-called Papapetrou procedure[21]: namely, by integrating the conservation equation over an infinitely extended space-like hypersurface Σ intersecting the "world-tube" of the body at a given time $t = $ const., and by expanding the gravitational field variables in power series around the world-line $x^\mu(t)$ of its center of mass. One obtains, in this way, a "multipole" expansion of the equation of motion including, at any given order, the gravitational coupling to all corresponding (dipole, quadrupole, etc.) internal momenta.

We are interested, in this chapter, in the case of a structureless, point-like test body. We can then neglect the contribution of all the internal momenta, and describe the test body with a delta-function distribution of its energy–momentum density, defined (see e.g. Ref. 22) by

$$T^{\mu\nu}(x') = \frac{1}{\sqrt{-g}}\delta^3(x' - x(t))\frac{p^\mu p^\nu}{p^0} \equiv \frac{m}{\sqrt{-g}}\delta^3(x' - x(t))\frac{\dot{x}^\mu \dot{x}^\nu}{\dot{x}^0}, \quad (6)$$

where $p^\mu = m\dot{x}^\mu$ and $p^0 = m\dot{x}^0 = mdt/d\tau$. In such a case, the volume integration over the space-like hypersurface Σ, namely $\int_\Sigma d^3x'\sqrt{-g}\nabla_\nu\theta^{\mu\nu}$, becomes trivial, and by applying the Gauss theorem to eliminate the integral of spatial divergences (there is no flux of $\theta^{\mu\nu}$ at spatial infinity), we obtain, from Eq. (5):

$$\frac{dp^\mu}{d\tau} + \Gamma_{\nu\alpha}{}^\mu\frac{p^\alpha p^\nu}{p^0} - \frac{2\alpha\pi}{3}\frac{d}{dt}\left(\frac{T^2}{E^2}V_4^\mu V_4^\nu\frac{p^\alpha p^\beta}{p^0}V_\alpha^4 V_\beta^4\right)$$

$$- \frac{2\alpha\pi}{3}\frac{T^2}{E^2}\Gamma_{\nu\alpha}{}^\mu V_4^\alpha V_4^\nu\frac{p^\beta p^\rho}{p^0}V_\beta^4 V_\rho^4 = 0. \quad (7)$$

Finally, let us express the time derivatives in terms of the proper time parameter τ, and multiply the above equation by $m^{-1}dt/d\tau = m^{-1}\dot{x}^0 = p^0/m^2$. By recalling that $p^\mu = m\dot{x}^\mu$, and by using for E the definition (4), we obtain

$$\ddot{x}^\mu + \Gamma_{\nu\alpha}{}^\mu\dot{x}^\alpha\dot{x}^\nu - \frac{2\alpha\pi}{3}\frac{T^2}{m^2}\frac{d}{d\tau}\left(V_4^\mu V_4^0\frac{m}{p^0}\right) - \frac{2\alpha\pi}{3}\frac{T^2}{m^2}\Gamma_{\nu\alpha}{}^\mu V_4^\alpha V_4^\nu = 0. \quad (8)$$

The last two terms describe the mass-dependent deviations from geodesic motion induced by the thermal corrections, to lowest order in T^2/m^2. A simple application of this equation, illustrating the non-universality of free-fall at finite temperature, will be presented in the next subsection.

2.1. Example: Radial motion in the Schwarzschild field

Let us consider the radial trajectory of test particle in the Schwarzschild geometry produced by a central source of mass M and described, in polar coordinates $x^\mu = (t, r, \theta, \phi)$ by the diagonal metric

$$g_{\mu\nu} dx^\mu dx^\nu = e^\nu dt^2 - e^{-\nu} dr^2 - r^2(d\theta^2 + \sin^2\theta d\phi^2), \quad e^\nu = 1 - 2GM/r. \tag{9}$$

The vierbein field is also diagonal, with

$$V_\mu^4 = \delta_\mu^4 e^{\nu/2}, \qquad\qquad V_4^\mu = \delta_4^\mu e^{-\nu/2}, \tag{10}$$

and the relevant components of the Christoffel connection, for a radial trajectory with $\dot\theta = 0$, $\dot\phi = 0$, are given by

$$\Gamma_{01}{}^0 = \frac{\nu'}{2}, \qquad \Gamma_{00}{}^1 = \frac{\nu'}{2} e^{2\nu}, \qquad \Gamma_{11}{}^1 = -\frac{\nu'}{2}, \tag{11}$$

where a prime denotes differentiation with respect to r.

The radial motion in this gravitational field, according to Eq. (8), is the described by the following two independent equations:

$$\ddot{t} + \dot\nu\dot{t} = 0, \qquad \ddot{r} + \frac{\nu'}{2}\left(e^{2\nu}\dot{t}^2 - \dot{r}^2 - \frac{2\alpha\pi}{3}\frac{T^2}{m^2}e^\nu\right) = 0, \tag{12}$$

whose integration (with the condition of vanishing radial velocity at spatial infinity, $\dot{r} \to 0$ for $r \to \infty$) gives

$$\dot{t} = e^{-\nu}, \qquad \dot{r}^2 = 1 - e^\nu + \frac{2\alpha\pi}{3}\frac{T^2}{m^2}\nu e^\nu. \tag{13}$$

By inserting this result into Eq. (12) we can finally write the generalized expression for the radial acceleration of a test particle at finite temperature as follows:

$$\ddot{r} = -\frac{GM}{r^2}\left\{1 - \frac{2\alpha\pi}{3}\frac{T^2}{m^2}\left[1 + \ln\left(1 - \frac{2GM}{r}\right)\right]\right\}. \tag{14}$$

This result describes, for $T > 0$, a non-universal, mass-dependent deviation from geodesic motion. The temperature-dependent corrections controlling the breaking of the equivalence principle are very small, however. In the weak field limit, in which terms higher than linear in the gravitational potential GM/r are neglected, we can estimate that the effective difference

Δm between inertial and gravitational mass, for a particle of rest mass m_0, is given by

$$\frac{|\Delta m|}{m_0} \sim \alpha \frac{T^2}{m_0^2}. \tag{15}$$

For macroscopic masses and ordinary values of the temperature this effect is well outside the present experimental sensitivities (see e.g. the results of the recent MICROSCOPE space mission[23]). Let us notice, for instance, that at a temperature $T \sim 300$ kelvin, and for an electron mass $m_0 \sim 0.5$ MeV, we have $\alpha (T/m_0)^2 \sim 10^{-18}$.

Assuming that the result (14) for the radial acceleration keeps valid if extrapolated to the strong gravity regime (i.e. at very small values of the radial coordinate), it may be interesting to note that the deviations from the geodesic trajectory are still mass depend, but the gravitational attraction tends to diverge, for *any* given value of m, when approaching the Schwarzschild radius $r = 2GM$ (as illustrated in Fig. 1).

Let us stress, however, that the result (14) is only valid in the limit $T/m \ll 1$. At higher temperatures, higher-order corrections to the particle

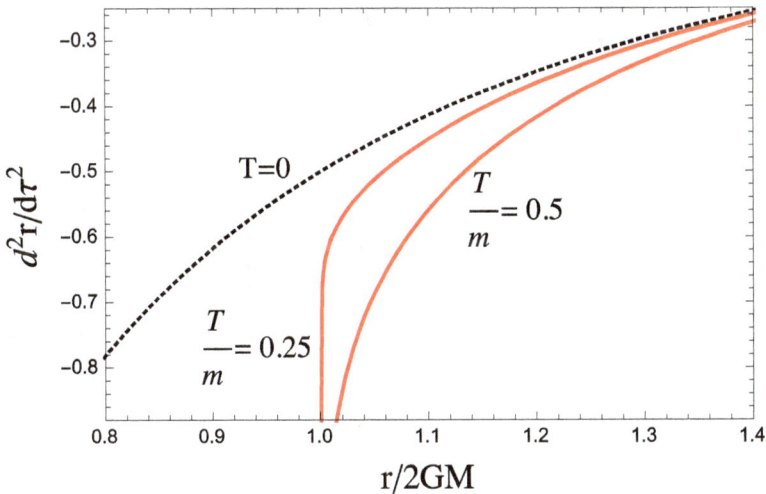

Fig. 1. The radial acceleration of Eq. (14) is plotted without temperature corrections ($T = 0$, black dashed curve) and with temperature corrections ($T > 0$, red solid curves). At finite temperature, the attractive force tends to diverge at $r = 2GM$ for any given value of the ratio $T/m < 1$.

trajectory (and possibly also to the effective space-time geometry, see Sec. 4) are needed.

3. Non-geodesic Motion for Locally Lorentz-Noninvariant Matter-Gravity Interactions

The energy–momentum tensor (2), modified by the thermal corrections, is only a particular example of matter distribution coupled to gravity with general-covariant but not locally Lorentz-invariant interactions.

More generally, assuming that the gravitational coupling to matter is only SO(3)-invariant in the local tangent space, we can decompose the standard energy–momentum $T^{\mu\nu}$ of the matter sources into its tangent space components T^{ij}, T^{i4} and T^{44} transforming, respectively, as a tensor, a vector, and a scalar under the local SO(3) group (conventions: Latin indices $i, j, k, \ldots$ run from 1 to 3). If the local Lorentz symmetry is broken, those different components may contribute with different coupling strength to the gravitational equations,[3–5] thus producing an effective source of gravity described by a modified energy–momentum tensor $\theta^{\mu\nu} = T^{\mu\nu} + \Delta T^{\mu\nu}$.

For the purpose of this chapter, we can conveniently (and equivalently) work with the SO(3)-scalar variables T^{44}, $T^{4\nu}$, $T^{\mu 4}$, so as to express the modified gravitational source in general-covariant form, and in terms of the minimally coupled matter stress tensor $T^{\alpha\beta}$, as follows[3–5]:

$$\theta^{\mu\nu} = T^{\mu\nu} + a_1 V_4^\mu V_4^\nu T^{44} + a_2 V_4^\mu T^{4\nu} + a_3 V_4^\nu T^{\mu 4}$$

$$\equiv T^{\mu\nu} + a_1 V_4^\mu V_4^\nu T^{\alpha\beta} V_\alpha^4 V_\beta^4 + a_2 V_4^\mu T^{\alpha\nu} V_\alpha^4 + a_3 V_4^\nu T^{\mu\alpha} V_\alpha^4. \quad (16)$$

Here a_1, a_2, a_3 are dimensionless parameters governing the breaking of the local SO(3, 1) symmetry, which is restored in the limit $a_1 = a_2 = a_3 = 0$.

It should be noted that the generalized tensor $\theta^{\mu\nu}$ is symmetric, $\theta^{\mu\nu} = \theta^{\nu\mu}$, provided that $T^{\mu\nu}$ is symmetric and $a_2 = a_3$. Note, also, that the finite-temperature stress tensor given by Eq. (2) can be exactly reproduced from Eq. (16) by putting $a_2 = a_3 = 0$ and $a_1 = -(2\alpha\pi/3)(T^2/E^2)$. In this section, we shall assume that the coefficients a_i are constant parameters; it should be stressed, however, that they might acquire an intrinsic energy dependence in a different (and probably more realistic) model of Lorentz symmetry breaking, as suggested indeed by the finite-temperature scenario discussed in Sec. 2.

Let us now assume, as in Sec. 2, that local Lorentz symmetry is broken only in the matter part of the action, in such a way that the generalized gravitational equations can be written as $G^{\mu\nu} = 8\pi G\theta^{\mu\nu}$, and the contracted Bianchi identity implies the conservation equation $\nabla_\nu \theta^{\mu\nu} = 0$. For consistency with the symmetry property of the Einstein tensor $G^{\mu\nu}$, we have to assume, of course, that $\theta^{\mu\nu}$ is also symmetric, which implies (as previously stressed) $T^{\mu\nu} = T^{\nu\mu}$ and $a_2 = a_3$. The conservation law of the energy–momentum tensor (16) thus provides the condition

$$\partial_\nu (\sqrt{-g}\, T^{\mu\nu}) + a_1\, \partial_\nu \left(\sqrt{-g}\, V_4^\mu V_4^\nu T^{\alpha\beta} V_\alpha^4 V_\beta^4\right)$$
$$+ a_2\, \partial_\nu \left(\sqrt{-g}\, V_4^\mu T^{\alpha\nu} V_\alpha^4 + \sqrt{-g}\, V_4^\nu T^{\mu\alpha} V_\alpha^4\right)$$
$$+ \sqrt{-g}\, \Gamma_{\nu\alpha}{}^\mu (T^{\alpha\nu} + a_1 V_4^\alpha V_4^\nu T^{\beta\rho} V_\beta^4 V_\rho^4 + a_2 V_4^\alpha T^{\beta\nu} V_\beta^4 + a_2 V_4^\nu T^{\alpha\beta} V_\beta^4)$$
$$= 0. \tag{17}$$

We shall follow the same procedure as in Sec. 2, by integrating the above equation over an infinitely extended spatial hypersurface Σ, by applying the Gauss theorem to eliminate the integral of the spatial divergences, and by assuming that $T^{\mu\nu}$ can be appropriately described by the delta-function distribution (6). By multiplying the result by $m^{-1} dt/d\tau$, we finally obtain the following generalized equation of motion:

$$\ddot{x}^\mu + \Gamma_{\nu\alpha}{}^\mu \dot{x}^\alpha \dot{x}^\nu + a_1 \frac{d}{d\tau}\left(V_4^\mu V_4^0 \frac{\dot{x}^\alpha \dot{x}^\beta}{\dot{x}^0} V_\alpha^4 V_\beta^4\right)$$
$$+ a_2 \frac{d}{d\tau}\left[\left(V_4^\mu + V_4^0 \frac{\dot{x}^\mu}{\dot{x}^0}\right) \dot{x}^\alpha V_\alpha^4\right] + a_1 \Gamma_{\nu\alpha}{}^\mu V_4^\alpha V_4^\nu \dot{x}^\beta \dot{x}^\rho V_\beta^4 V_\rho^4$$
$$+ a_2 \Gamma_{\nu\alpha}{}^\mu \left(V_4^\alpha \dot{x}^\nu + V_4^\nu \dot{x}^\alpha\right) \dot{x}^\beta V_\beta^4 = 0, \tag{18}$$

which describes the non-geodesic trajectory of a point-like test body coupled to gravity in a way which preserves the local rotational symmetry, but breaks in general the local Lorentz invariance. The geodesic deviations are controlled by the Lorentz-breaking parameters a_1 and a_2.

In the following subsection, we shall apply this result to discuss the possible effects on the radial acceleration of a free-falling test body in the static, spherically symmetric field of a central source.

3.1. *Example: Repulsive forces and possible gravitational non-universality*

Let us consider, as in Sec. 2.1, a radial motion in the Schwarzschild geometry described by the metric (9). The relevant components of the vierbein field and of the Christoffel connection are given by Eqs. (10) and (11). By writing explicitly the $\mu = 0$ and $\mu = 1$ components of the equation of motion (18) we obtain, respectively, the following two independent equations[4,5]:

$$(1 + a_2 + 2a_2)\, \ddot{t} + (1 + a_2)\, \nu' \dot{r} \dot{t} = 0, \tag{19}$$

$$(1 + a_2)\, \ddot{r} - \frac{\nu'}{2} \dot{r}^2 + (1 + a_1 + 2a_2)\, \frac{\nu'}{2} e^{2\nu} \dot{t}^2 = 0. \tag{20}$$

Assuming that $1 + a_1 + 2a_2 \neq 0$ and $1 + a_2 \neq 0$ (we expect indeed that all Lorentz-breaking corrections are small, $|a_1| \ll 1$, $|a_2| \ll 1$), we can define the convenient parameters

$$\beta = \frac{1 + a_2}{1 + a_1 + 2a_2}, \qquad \gamma = \frac{1}{1 + a_2}, \tag{21}$$

and we can express the integration of Eqs. (19) and (20) (with the initial condition of vanishing radial velocity at spatial infinity) as follows:

$$\dot{t} = e^{-\beta \nu}, \qquad \dot{r}^2 = \frac{1}{\beta(2 - \gamma - 2\beta)} [e^{\gamma \nu} - e^{2\nu(1-\beta)}]. \tag{22}$$

Finally, by inserting this result into Eq. (20), we find that the generalized radial acceleration of the test body is given by

$$\ddot{r} = -\frac{GM}{r^2} \frac{1}{\beta(2 - \gamma - 2\beta)}$$
$$\times \left[2(1 - \beta) \left(1 - \frac{2GM}{r} \right)^{1-2\beta} - \gamma \left(1 - \frac{2GM}{r} \right)^{\gamma - 1} \right]. \tag{23}$$

For $\beta = \gamma = 1$ we recover the standard result $\ddot{r} = -GM/r^2$. Note that the above expression is different from the modified trajectory described by Eq. (14), because at finite temperature the Lorentz-breaking terms are controlled by the energy-dependent (and thus time-dependent) coefficient T^2/E^2, which provides additional contributions to the geodesic deviations through its non-vanishing time derivative (see Eq. (7)). In this section, instead, we have assumed constant Lorentz-breaking parameters, $\dot{a}_1 = 0$, $\dot{a}_2 = 0$.

It may be interesting to note however that, even with such a simplified model, and for an appropriate choice of the Lorentz-breaking parameters, we may expect the automatic appearance of repulsive gravitational interactions. This occurs (for instance) if $1/2 < \beta < 1$ and $\beta > 1 - \gamma/2$; and the last condition, incidentally, is automatically satisfied if the motion has to preserve causality in the spatial region $r > 2GM$ (see Ref. 4 for a detailed

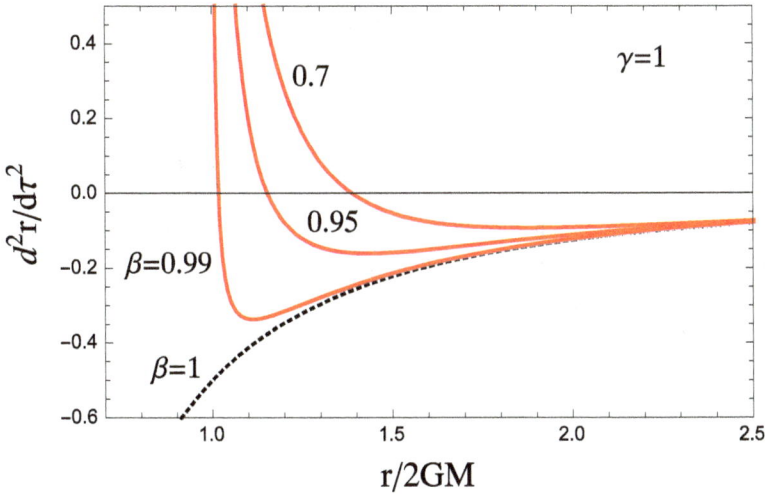

Fig. 2.   The radial acceleration of Eq. (23) is plotted for $\gamma = 1$ and for different values of β. For $\beta = 1$, we have the standard result of general relativity (black dashed curve). For $\beta < 1$, we have repulsive gravitational interactions which become dominant at small enough values of r, and diverge at $r = 2GM$ (red solid curves with $\beta = 0.99$, $\beta = 0.95$ and $\beta = 0.7$).

discussion of the allowed numerical values of β and γ in order to avoid the presence of imaginary radial velocities and space-like four-velocity vectors).

The repulsive interactions, when present, become dominant at small enough radial distances, and tend to diverge in the limit $r \to 2GM$, as illustrated in Fig. 2 for particular values of β and γ. In that case $\dot{r} \to 0$ at $r = 2GM$, and the interior of the Schwarzschild sphere becomes a "classically impenetrable" region[4] (an effect similar to the one occurring in the context of Rosen bimetric theory of gravity[24]).

Let us stress that in the model we are considering the deviations from geodesic motion are triggered by the Lorentz-breaking parameters a_1, a_2 which are not necessarily mass dependent (unlike the finite-temperature corrections discussed in Sec. 2). If so, the resulting free motion of a test particle in a given gravitational field is non-geodesic, but still "universal".

In principle, however, the effective violation of the local Lorentz symmetry might be different for different types of particles, thus producing an effective non-universality of free fall and of the gravitational coupling,[5, 25, 26] which could be tested by applying the generalized equation of motion (23).

For instance, let us assume (as a working hypothesis) that the local $SO(3,1)$ symmetry is broken for the gravitational interactions of baryons but not for those of leptons.[5] This clearly produces a "composition-dependent" violation of the equivalence principle, very similar, in practice, to that produced by the coupling of the so-called *fifth force* to the baryon number[27, 28] (see also the recent discussion of Ref. 29). In such a case, if we have a macroscopic test body of mass m_1 containing B_1 baryons of mass m_B, and if we consider Eq. (23) in the weak field limit (neglecting terms higher than linear in the gravitational potential GM/r), we find that the effective gravitational force acting on m_1 is given by

$$m_1 \ddot{r}_1 = -\frac{GM}{r^2} \frac{m_1}{\beta} = -\frac{GM}{r^2} m_1 \left[1 + \left(\frac{a_1 + a_2}{1 + a_2} \right) \frac{m_B}{m_1} B \right] \qquad (24)$$

(we have set $m_1 = m_B B$).

Comparing the accelerations $\ddot{r}_1$ and $\ddot{r}_2$ of two different test masses m_1 and m_2, in the Earth (or solar) gravitational field, we thus obtain

$$\frac{\Delta a}{g} = \left(\frac{a_1 + a_2}{1 + a_2} \right) \Delta \left(\frac{B}{\mu} \right), \qquad (25)$$

where $\Delta a = \ddot{r}_1 - \ddot{r}_2$, where $g = -GM/r^2$ is the local acceleration of gravity, and where $\Delta(B/\mu) = (B_1/\mu_1) - (B_2/\mu_2)$, with $\mu = m/m_B$ the mass of the test body in units of baryonic mass. Hence, a different violation of the local Lorentz symmetry for baryons and leptons leads to a composition-dependent gravitational acceleration of macroscopic test bodies, which — for constant values of a_1 and a_2 — is strongly constrained by existing experimental results.

According to the most recent tests of the equivalence principle,[23] we can impose, in fact, the upper limit $(\Delta a/g) \lesssim 10^{-15}$ for bodies with $\Delta(B/\mu) \sim 10^{-3}$. This implies

$$\left| \frac{a_1 + a_2}{1 + a_2} \right| \lesssim 10^{-12} \qquad (26)$$

(unless, of course, we consider more sophisticated models of Lorentz symmetry breaking where the constant parameters a_1, a_2 are replaced by position-dependent and/or energy-dependent variables).

4. A Quasi-Riemannian Model of Gravity with Local SO(3) Tangent-Space Symmetry

We have shown, in the previous sections, that a breaking of the local $SO(3,1)$ symmetry — like that occurring at finite temperature — leads to modify the coupling of the test bodies to the background geometry. In such a context, it may be natural to expect a modified dynamics also for the geometry itself: in particular, a dynamics described by gravitational equations which are not locally Lorentz-invariant but only SO(3)-invariant.

A simple way to formulate an effective theory of this type is to follow the scheme of the so-called *quasi-Riemannian* models of gravity,[6–9] and to choose, in particular the rotational group SO(3) as the dynamical gauge symmetry of the flat space-time locally tangent to the (curved) world manifold.[3–5]

It is convenient, to this purpose, to construct the action working directly in the local tangent space, where the Lorentz connection $\omega_\mu{}^{ab}$ can be decomposed into the SO(3) connection $\omega_\mu{}^{ij}$ and the SO(3) vector $\omega_\mu{}^{i4}$ (let us recall that $i, j, k, \ldots$ run from 1 to 3). Using these variables, plus the local components of the vierbein field V_μ^i, V_μ^4 (transforming, respectively, as an SO(3) vector and scalar field), we can then easily write a modified gravitational action which is generally covariant but, locally, only SO(3) invariant.

It may be useful, also, to adopt the compact language of differential forms, and work with the connection one-form $\omega^{ab} \equiv \omega_\mu{}^{ab} dx^\mu$ and the anholonomic basis one-form $V^a \equiv V_\mu^a dx^\mu$. In this formalism, the standard Einstein action can be written in terms of the curvature two-form R^{ab} as

$$S_E = -\int d^4x \sqrt{-g}\, R \equiv \frac{1}{2}\int R^{ab} \wedge V^c \wedge V^d \epsilon_{abcd},$$

$$R^{ab} = d\omega^{ab} + \omega^a{}_c \wedge \omega^{cd}. \tag{27}$$

Conventions: the symbol "*d*" denotes exterior derivative, and the wedge symbol "$\wedge$" exterior product; finally, ϵ is the totally antisymmetric Levi-Civita symbol of the flat tangent space.

We can now introduce the possible Lorentz breaking — but SO(3) preserving — contributions, and write the generalized gravitational action in quasi-Riemannian form as follows:

$$S = \frac{1}{16\pi G}\int \left[\frac{1}{2}\int R^{ab} \wedge V^c \wedge V^d \epsilon_{abcd} + \left(b_1\, \overline{R}^{ij} \wedge V^k \wedge V^4\right.\right.$$

$$\left.\left. + b_2\, \overline{D}\omega^{i4} \wedge V^j \wedge V^k + b_3\, \omega^i{}_4 \wedge \omega^{j4} \wedge V^k \wedge V^4\right)\epsilon_{ijk4} + \cdots\right], \tag{28}$$

where we have explicitly introduced the SO(3) curvature (or Yang–Mills) term $\overline{R}^{ij}$ and the SO(3) covariant (exterior) derivative $\overline{D}$, defined by

$$\overline{R}^{ij} = d\omega^{ij} + \omega^i{}_k \wedge \omega^{kj}, \qquad \overline{D}\omega^{i4} = d\omega^{i4} + \omega^i{}_k \wedge \omega^{k4}. \qquad (29)$$

The dimensionless coefficients b_i are constant parameters controlling the breaking of the local Lorentz symmetry, and the stand gravitational theory is recovered in the limit $b_i = 0$. Finally, the dots denote the possible addition of other SO(3)-invariant contributions, that may be present or not depending on the chosen model of Lorentz-symmetry braking, as well as on the assumed type of geometry (e.g., with or without torsion, nonmetricity tensor, and so on). See Refs. 3 and 11 for a general discussion.

By adding the action for the matter sources, and by varying the total action with respect to the ω^{ab} and V^a we then obtain, respectively, the explicit expression for the connection and the generalized form of the gravitational equations (see Refs. 3, 7 and 9 for detailed computations). The final result for the modified Einstein equations can be written in general as

$$G^{\mu\nu} + \Delta G^{\mu\nu} = 8\pi G\theta^{\mu\nu} \equiv 8\pi G \left(T^{\mu\nu} + \Delta T^{\mu\nu} \right), \qquad (30)$$

where the right-hand side of these equations exactly corresponds to the generalized matter stress tensor $\theta^{\mu\nu}$ of Eq. (16). On the left-hand side, we have the usual Einstein tensor, $G^{\mu\nu} = R^{\mu\nu} - Rg^{\mu\nu}/2$, plus the corrections $\Delta G^{\mu\nu}$ induced by the breaking of the local Lorentz symmetry.

It should be noted that $G^{\mu\nu}$ is a symmetric tensor, but $\Delta G^{\mu\nu} \neq \Delta G^{\nu\mu}$, in general. Hence, there is no need of imposing on $\Delta T^{\mu\nu}$ to be symmetric, and we may have $a_2 \neq a_3$ in Eq. (16).

Note also that the contracted Bianchi identity $\nabla_\nu G^{\mu\nu} = 0$ leads to the conditions

$$\nabla_\nu T^{\mu\nu} = \nabla_\nu \left(\frac{\Delta G^{\mu\nu}}{8\pi G} - \Delta T^{\mu\nu} \right), \qquad (31)$$

which implies, in general, deviations from the geodesic motion of free-falling test particles (as discussed in the previous sections). However, given that our modified gravitational equations depend on 6 (or more) parameters, $a_1, a_2, a_3, b_1, b_2, b_3, \ldots$, it turns out that it is always possible in principle to preserve a geodesic type of motion ($\nabla_\nu T^{\mu\nu} = 0$) by imposing as a constraint that the right-hand side of Eq. (31) is identically vanishing. This constraint provides indeed four additional conditions which reduce the number of independent parameters for this class of models (see Sec. 4.1 for an explicit example of this possibility).

For the illustrative purpose of this chapter, we shall concentrate on a simple model of quasi-Riemannian gravity where the breaking of the local Lorentz symmetry leads to modified equations which can be written in terms of the Ricci tensor R_μ^ν as follows:

$$R_\mu{}^\nu + \Delta R_\mu{}^\nu = 8\pi G \left(\theta_\mu{}^\nu - \frac{1}{2} \delta_\mu^\nu \theta \right), \tag{32}$$

$$\Delta R_\mu{}^\nu = R_\mu{}^\nu + c_1 R_{\mu 4}{}^{\nu 4} + c_2 V_\mu^4 V_\nu^\nu R_4{}^4 + c_3 V_\mu^4 R_4{}^\nu + c_4 \omega_{\mu \alpha 4} \omega^{\nu \alpha 4}$$

$$= \left(c_1 R_{\mu \alpha}{}^{\nu \beta} + c_2 V_\mu^4 V_4^\nu R_\alpha{}^\beta \right) V_4^\alpha V_\beta^4 + c_3 V_4^4 R_\alpha{}^\nu V_4^\alpha$$

$$+ c_4 \omega_{\mu \alpha \beta} \omega^{\nu \alpha \rho} V_4^\beta V_\rho^4. \tag{33}$$

Here $\theta_\mu{}^\nu$ is given by Eq. (16), $\theta = \theta_\alpha{}^\alpha$, $R_{\mu \alpha}{}^{\nu \beta}$ is the usual Riemann tensor, and $c_1, \dots, c_4$ are the constant Lorentz-breaking parameters. Finally, the tangent space connection ω is fixed as usual by the so-called *metricity postulate*,

$$\nabla_\mu V_\nu^a = \partial_\mu V_\nu^a + \omega_\mu{}^a{}_b V_\nu^b - \Gamma_{\mu \nu}{}^\alpha V_\alpha^a = 0. \tag{34}$$

In the following subsection, we shall apply the above equations to describe the cosmological geometry produced by a distribution of perfect-fluid matter sources.

4.1. *Example: Cosmological applications with and without violation of the equivalence principle*

Let us consider the spatially homogeneous and isotropic Friedmann–Lemaitre–Robertson–Walker (FLRW) geometry, described in polar coordinates $x^\mu = (t, r, \theta, \phi)$ by the metric

$$ds^2 = g_{\mu\nu} dx^\mu dx^\nu = dt^2 - a^2(t) \left[\frac{dr^2}{1 - kr^2} + r^2 d\theta^2 + r^2 \sin^2 \theta d\phi^2 \right], \tag{35}$$

where t is the cosmic time, $a(t)$ the scale factor, and $k = 0, \pm 1$ the constant spatial curvature. The unperturbed energy–momentum $T^{\mu\nu}$ of the fluid sources, assumed at rest in the comoving frame, is given by the diagonal tensor

$$T_\mu{}^\nu = \mathrm{diag}\,(\rho, -p, -p, -p), \tag{36}$$

where the (time-dependent) energy density ρ and pressure p are related by a barotropic equation of state, $p/\rho = w = \mathrm{const}$.

For this geometry, we simply have $V_\mu^4 = \delta_\mu^4$, $V_4^\mu = \delta_4^\mu$, and the relevant components of the tangent space connection ω, fixed by Eq. (34), are given by $\omega_\mu{}^\alpha{}_4 = \Gamma_{\mu 4}{}^\alpha = H(\delta_\mu^1 \delta_1^\alpha + \delta_\mu^2 \delta_2^\alpha + \delta_\mu^3 \delta_3^\alpha)$, where $H = \dot{a}/a$ (the dot denotes differentiation with respect to the cosmic time t). The generalized equations (32) reduce, in this case, to the following two independent equations,

$$-3\frac{\ddot{a}}{a}(1 + c_2 + c_3) = 4\pi G\left[\rho(1 + a_1 + a_2 + a_3) + 3p\right], \qquad (37)$$

$$\frac{\ddot{a}}{a}(1 - c_1) + 2H^2\left(1 - \frac{c_4}{2}\right) + \frac{2k}{a^2} = 4\pi G\left[\rho(1 + a_1 + a_2 + a_3) - p\right], \qquad (38)$$

obtained , respectively, from the $(0,0)$ and $(1,1)$ components of of Eq. (32). In the limit $c_i = 0$, $a_i = 0$, we exactly recover the Einstein equations for the metric (35) and the matter distribution (36).

Let us now consider two simple particular cases, describing "minimal" (but interesting) modifications of the standard cosmological scenario.

The first one is based on the assumption that the Lorentz-breaking corrections lead to a new, modified cosmological dynamics which leaves unchanged, however, the form of the well-known Friedmann equation.[11] Such a scenario can be obtained, in the context of our model, by the following choice of parameters:

$$a_1 = a_2 = a_3 = 0, \qquad -c_1 = c_2 + c_3 \neq 0, \qquad c_4 = 0. \qquad (39)$$

With that choice, in fact, by eliminating $\ddot{a}/a$ from Eq. (37) in terms of Eq. (38), and using the identity $\ddot{a}/a = \dot{H} + H^2$, we obtain that the two modified cosmological equations can be rewritten, respectively, as

$$H^2 + \frac{k}{a^2} = \frac{8\pi G}{3}\rho, \qquad (40)$$

$$2\dot{H}(1 - c_1) + 3H^2\left(1 - \frac{2}{3}c_1\right) + \frac{k}{a^2} = -8\pi Gp, \qquad (41)$$

and that their combination gives

$$\dot{\rho}(1 - c_1) + 3H(\rho + p) = 2c_1 H\rho. \qquad (42)$$

We are thus left with an unchanged Friedmann equation (40), but we have a corresponding non-trivial modification of the spatial Einstein equation (41) and of the covariant evolution in time of the energy–momentum density, (42) (which is no longer equivalent to the conservation law $\nabla_\nu T^{\mu\nu} = 0$).

As discussed in Ref. 11, such a minimal, one-parameter-dependent violation of the local Lorentz symmetry may have interesting applications in a primordial cosmological context, where — if the violation is strong enough — it can produce accelerated (inflationary) expansion even in the absence of exotic sources with negative pressure (like, for instance, an effective cosmological constant).

Consider in fact an early enough epoch when the Universe is still radiation dominated ($p = \rho/3$), and the contribution of the spatial curvature to the cosmic dynamics is negligible, so that we can put $k = 0$ in Eqs. (40)–(42). A simple integration of those equations then gives

$$\rho(t) \sim a^{-(4-2c_1)/(1-c_1)}, \quad a(t) \sim t^{(1-c_1)/(2-c_1)}. \tag{43}$$

For $c_1 > 2$, this solution describes a phase of accelerated expansion of "power-law" type, with $\ddot{a} > 0$ and $\dot{H} < 0$. In the limit $c_1 \to 2$, the solution describes a phase of exponential inflation, $a \to \exp(Ht)$, $\dot{H} \to 0$ (with no need of introducing, to this purpose, an effective inflaton field assumed to be "slow rolling" along some *ad hoc* inflaton potential).

Let us now report the second example of modified cosmological dynamics where, in spite of the corrections due to the Lorentz-breaking terms, the covariant conservation of the standard energy–momentum tensor is preserved, $\nabla_\nu T^{\mu\nu} = 0$, and the evolution in time of the matter sources is geodesic.[3, 12] This possibility corresponds to a model with the following values of the parameters:

$$a_1 = a_2 = a_3 = 0, \qquad -c_1 = c_2 + c_3 \neq 0, \qquad c_4 = 2c_1 \neq 0. \tag{44}$$

In that case, by eliminating $\ddot{a}/a$ from Eq. (37) in terms of Eq. (38), we find that Eqs. (37) and (38) can be rewritten as

$$H^2(1 - c_1) + \frac{k}{a^2} = \frac{8\pi G}{3}\rho, \tag{45}$$

$$2\dot{H}(1 - c_1) + 3H^2(1 - c_1) + \frac{k}{a^2} = -8\pi Gp, \tag{46}$$

and that their combination gives

$$\dot{\rho} + 3H(\rho + p) = 0, \tag{47}$$

which exactly corresponds to the standard conservation law of the energy–momentum (36).

Note that in the absence of spatial curvature, $k = 0$, this particular model of Lorentz symmetry breaking has no dynamical effects on the evolution of the cosmic geometry apart from a trivial renormalization of the coupling constant, $G \to G/(1 - c_1)$. In that case, for $c_1 > 1$, we would find always repulsive gravitational interactions, a possibility which is clearly excluded by standard gravitational phenomenology.

Interestingly enough, however, if $k > 0$ (and $c_1 > 1$), the repulsive interactions may become dominant in the limit of high enough energy density (i.e. during the very early cosmological phases), and allow non-singular "bouncing" solutions to the cosmological equations even in the presence of conventional sources satisfying the strong energy condition.[3, 12] This is possible because, according to the modified equations (45)–(47), the condition which makes the singularity unavoidable (i.e. the condition of geodesic convergence $R_{\mu\nu} u^\mu u^\nu \geq 0$, where u^μ is a time-like vector field), and the strong energy condition, $T_{\mu\nu} - g_{\mu\nu} T/2 \geq 0$, are no longer equivalent conditions.

To give an explicit example let us consider a radiation fluid with $p = \rho/3$ and a cosmic geometry with $k = 1/t_0^2 > 0$, where $t_0 = $ const. is a given parameter controlling the spatial curvature scale. From Eq. (47) we obtain $\rho = \rho_0 a^{-4}$, where ρ_0 is an integration constant, and the modified Friedmann equation (45), with $c_1 > 1$, has the particular exact solution

$$a(t) = \left[\frac{8\pi G}{3} \rho_0 t_0^2 + \frac{(t/t_0)^2}{|1 - c_1|} \right]^{1/2}, \qquad (48)$$

with the cosmic time t ranging from $-\infty$ to $+\infty$. The associated Hubble parameter is given by

$$H = \frac{t/t_0^2}{(t/t_0)^2 + |1 - c_1| 8\pi G \rho_0 t_0^2/3}, \qquad (49)$$

and has no singularity in the whole range $-\infty \leq t \leq +\infty$.

As illustrated in Fig. 3, the above solution describes a continuous and regular bouncing transition between two complementary (or "dual") cosmological phases,[a] defined, respectively, in the time ranges $t < 0$ and $t > 0$. The initial, asymptotically flat regime describes, for $t < 0$, a collapsing phase of decelerated contraction ($\dot{a} < 0$, $\ddot{a} > 0$), initially growing (in modulus) curvature scale ($\dot{H} < 0$), and growing energy density ($\dot{\rho} > 0$). The density reaches the maximum value ρ_0 at the epoch $t = 0$, which marks a smooth transition towards the final regime characterized, for $t > 0$ by accelerated expansion ($\dot{a} > 0$, $\ddot{a} > 0$), eventually decreasing curvature scale ($\dot{H} < 0$), and decreasing energy density ($\dot{\rho} < 0$).

Let us stress, finally, that the two examples of modified gravitational dynamics reported in this section require, for an efficient application in a

---

[a]See Refs. 14–16, 30 and 31 for similar scenarios in a string cosmology context.

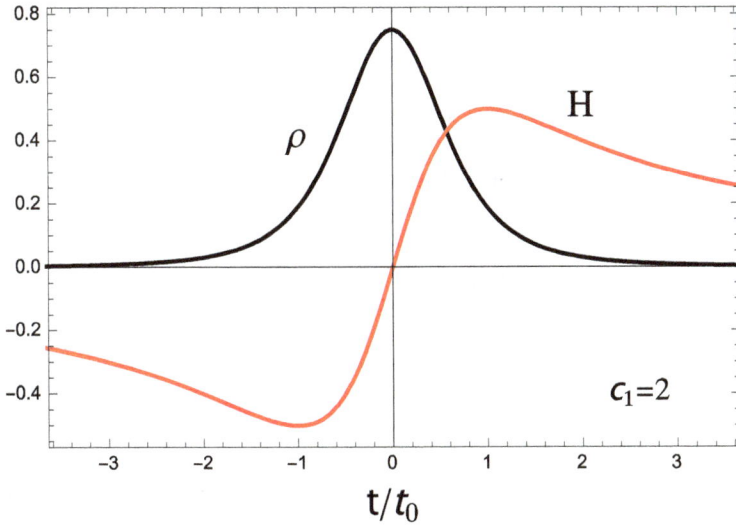

Fig. 3. Smooth evolution of the radiation energy density $\rho = \rho_0/a^4$ (black solid curve) and of the Hubble parameter H (red solid curve) though the bouncing transition described by the solution (48) and (49). The curves are plotted for $c_1 = 2$, $8\pi G \rho_0 t_0^2/3 = 1$, $\rho_0 = 0.75$, and H is expressed in units of $1/t_0$.

cosmological context, relatively large values of the Lorentz-breaking parameters ($|c_i| \sim 1$). Hence, they are expected to possibly describe a realistic scenario only at very early epochs, when the Universe approaches the Planck energy scale and/or the quantum gravity regime. We know indeed, from many experimental data, that at lower energies the local Lorentz symmetry of the gravitational interactions is very efficiently restored, and only very weak violations are possibly allowed.

5. Conclusion

The principle of equivalence is at the very ground of Einstein's theory of gravity. According to this principle, the gravitational interaction can always be locally eliminated, and we can always locally reduce to the physics of the flat space-time, governed by the principle of Lorentz invariance.

If the effective local Lorentz invariance is broken (for instance, due to the presence of a thermal bath at finite temperature), we can then expect violations of the equivalence principle. Conversely, violations of such a principle, and deviations from the geodesic motion of point-like test bodies, may correspond to a local symmetry different from the Lorentz one.

The Lorentz symmetry group, on the other hand, is the gauge group of the General Relativity theory of gravity (where the curvature tensor plays the role of the non-Abelian "Yang–Mills field" of the local $SO(3,1)$ symmetry). If we adopt a different gauge group we can formulate a different gravitational theory, where the space-time geometry is still described in terms of Riemannian manifolds, but with a dynamics controlled by field equations different from Einstein's equations (see also Ref. 32 for a recent discussion of the independence of general coordinate transformations and local Lorentz rotations). In this chapter, we have considered, in particular, a possibly modified gravitational dynamics based on the local gauge group of the spatial rotations.

In that case, the principle of equivalence is not satisfied, in general, unless we impose appropriate constraints on the chosen model of Lorentz-symmetry breaking. In addition, the breaking can also produce gravitational interactions of repulsive type, which may have a relevant impact on the primordial cosmological dynamics.

On a macroscopic scale of energies and distances, however, we know that the possible violations of the local Lorentz symmetry and of the equivalence principle are both constrained by present experimental data to be extremely weak, and to produce only subdominant effects. Nevertheless, we believe that the possibility of such effects should be included when studying models and applications of the gravitational interaction at very high energies and in the quantum regime.

Acknowledgments

This work is supported in part by INFN under the program TAsP (*Theoretical Astroparticle Physics*), and by the research grant number 2017W4HA7S (*NAT-NET: Neutrino and Astroparticle Theory Network*), under the program PRIN 2017, funded by the Italian Ministero dell'Università e della Ricerca (MUR).

References

1. M. Gasperini, *Phys. Rev. D.* **36**, 617 (1987).
2. M. Gasperini, *Class. Quantum Grav.* **5**, 521 (1988).
3. M. Gasperini, *Class. Quantum Grav.* **4**, 485 (1987).
4. M. Gasperini and A. Tartaglia, *Mod. Phys. Lett. A.* **6**, 385 (1987).
5. M. Gasperini, Lorentz noninvariance and the universality of free fall in quasi-Riemannian gravity, In eds. V. De Sabbata and V. N. Melnikov, *Gravitational*

Measurements, Fundamental Metrology and Constants, pp. 181–190. Kluwer Academic Publishers (1988).

6. S. Weinberg, *Phys. Lett. B.* **138**, 47 (1984).
7. S. P. De Alwis and S. Randjbar-Daemi, *Phys. Rev. D.* **32**, 1345 (1985).
8. K. S. Viswanathan and B. Wong, *Phys. Rev. D.* **32**, 3108 (1985).
9. M. Gasperini, *Phys. Rev. D.* **33**, 3594 (1986).
10. V. De Sabbata and M. Gasperini, Gravitation without Lorentz invariance, In eds. P. G. Bergmann and V. De Sabbata, *Topological Properties and Global Structure of Spacetime*, p. 231. Plenum Publishing Corporation, New York (1986).
11. M. Gasperini, *Phys. Lett. B.* **163**, 84 (1985).
12. M. Gasperini, *Gen. Rel. Grav.* **30**, 1703 (1998).
13. M. Gasperini, *Phys. Rev. D.* **34**, 2260 (1986).
14. M. Gasperini, *Elements of String Cosmology*, Cambridge University Press, Cambridge, UK (2007).
15. M. Gasperini and G. Veneziano, *Nuovo Cim. C.* **38**, 160 (2016).
16. M. Gasperini, Elementary introduction to pre-big bang cosmology and to the relic graviton background, In eds. I. Ciufolini, V. Gorini, U. Moschella and P. Fre', *Gravitational Waves*, pp. 280–337. IOP Publishing, Bristol (2001).
17. M. Gasperini, *Phys. Rev. D.* **64**, 043510 (2001).
18. F. Donoghue, B. R. Holstein and R. W. Robinett, *Phys. Rev. D.* **30**, 2561 (1984).
19. F. Donoghue, B. R. Holstein and R. W. Robinett, *Gen. Rel. Grav.* **17**, 207 (1985).
20. F. Donoghue, B. R. Holstein and R. W. Robinett, *Phys. Rev. D.* **34**, 1208 (1986).
21. A. Papapetrou, *Proc. Roy. Soc. A.* **209**, 248 (1951).
22. S. Weinberg, *Gravitation and Cosmology*, Wiley, New York (1972).
23. P. Touboul *et al.*, *Class. Quantum Grav.* **36**, 225006 (2019).
24. N. Rosen, The space-time of the bimetric general relativity theory, In eds. P. G. Bergmann and V. De Sabbata, *Topological Properties and Global Structure of Spacetime*, p. 77. Plenum Pub. Co., New York (1986).
25. V. De Sabbata and M. Gasperini, Testing the local $SO(3, 1)$ invariance of gravity at the planetary level, In ed. R. Ruffini, *Proc. Fourth Marcel Grossmann Meeting on General Relativity*, p. 1709. Elsevier Science Publishers (1986).
26. V. De Sabbata and M. Gasperini, *Nuovo Cim. A.* **65**, 479 (1981).
27. E. Fischbach *et al.*, *Phys. Rev. Lett.* **56**, 3 (1986).
28. C. Talmadge and E. Fishchbach, *Searching for the source of the Fifth Force*, In eds. V. De Sabbata and V. N. Melnikov, *Gravitational Measurements, Fundamental Metrology and Constants*, p. 143. Kluwer Academic Publishers (1988).
29. E. Fischbach *et al.*, arXiv: 2012.2862 (2020).
30. M. Gasperini and G. Veneziano, *Astropart. Phys.* **1**, 317 (1993).
31. M. Gasperini and G. Veneziano, *Phys. Rept.* **373**, 1 (2003).
32. K. Cahill, *Phys. Rev. D.* **102**, 065011 (2020).

Chapter 4

An Information Theoretic Approach to the Weak Equivalence Principle

James Q. Quach[*,‡‡,***], Mir Faizal[†,‡,§,§§], Richard A. Norte[¶,‖,¶¶],
and Sebastian Bahamonde[**,††,‖‖]

[*]Institute for Photonics and Advanced Sensing and
School of Physical Sciences, The University of Adelaide,
South Australia 5005, Australia
[†]Department of Physics and Astronomy, University of Lethbridge,
4401 University Drive, Lethbridge, Alberta T1K 3M4, Canada
[‡]Irving K. Barber School of Arts and Sciences,
University of British Columbia — Okanagan, 3333 University Way,
Kelowna, British Columbia V1V 1V7, Canada
[§]Canadian Quantum Research Center, 204-3002, 32 Ave, Vernon,
BC, V1T 2L7, Canada
[¶]Department of Precision and Microsystems Engineering,
Delft University of Technology, Mekelweg 2, 2628CD Delft,
The Netherlands
[‖]Kavli Institute of Nanoscience, Delft University of Technology,
Lorentzweg 1, 2628CJ Delft, The Netherlands
[**]Department of Physics, Tokyo Institute of Technology
1-12-1 Ookayama, Meguro-ku, Tokyo 152-8551, Japan
[††]Laboratory of Theoretical Physics, Institute of Physics,
University of Tartu, W. Ostwaldi 1, 50411 Tartu, Estonia
[***]Commonwealth Scientific and Industrial Research Organization (CSIRO),
Clayton, Victoria, 3168, Australia
[‡‡]quach.james@gmail.com
[§§]mirfaizalmir@googlemail.com
[¶¶]r.a.norte@tudelft.nl
[‖‖]bahamonde.s.aa@m.titech.ac.jp

For the weak equivalence principle (WEP) to hold, we should not be able to gain any information about mass from its interaction with gravitational fields. This motivates the use of information theoretic techniques to investigate WEP violation. Using this approach, we demonstrate that the WEP holds for a quantum particle in a uniform gravitational field, but is violated in non-uniform and time-dependent gravitational fields, such as in gravitational waves. This provides a precise characterization of WEP violation by quantum systems in gravitational fields, that should be useful in formalizing other works that have argued for such violations heuristically. In particular, we discuss the possibility of detecting the gravitational Casimir effect with superconductors from an information theoretic perspective.

1. Introduction

The weak equivalence principle (WEP) states that point particles in free-fall will follow trajectories that are independent of their mass. This principle underpins classical gravitational theory, and is related to the assumption that the gravitational mass is equivalent to the inertial mass. In the context of classical theory, the WEP is well-defined; in quantum theory however, the WEP is ill-defined. This is because under Heisenberg's uncertainty principle, point particles and trajectories are ambiguous concepts. The problem is further highlighted when one compares the classical action of a field in gravity with is quantum counterpart. The classical action is

$$S_\mathrm{C} = -mc \int ds, \tag{1}$$

where $ds^2 = g_{\mu\nu}dx^\mu dx^\nu$. We use the Einstein summation convention where the repeated indices $\mu, \nu \in \{0, 1, 2, 3\}$ are summed. As m appears simply as a multiplicative factor, it does not feature in the equations of motion. This is consistent with the WEP. Implicit in this equation is the equivalence of the inertial and gravitational mass. In comparison, the quantum action is

$$S_\mathrm{Q} = \frac{\hbar^2}{2m} \int \sqrt{-g}\left(g^{\mu\nu}D_\mu\Phi^\dagger D_\nu\Phi + \frac{m^2c^2}{\hbar^2}\Phi^\dagger\Phi\right)\mathrm{d}^4x, \tag{2}$$

where D_μ is the covariant derivative in curved space-time. In this case, m simply is not a multiplicative factor, and features in the Klein–Gordon equation.

The WEP's notion that free-falling trajectories should be independent of mass can be reformulated as the statement that the Fisher information of a free-falling object is invariant with changes in mass.[1,2] In this information-theoretic framework, violation of the WEP means that one may extract information about an object's mass in free-fall. This information-theoretic formulation of the WEP has the advantage that it is extendable to quantum objects in an unambiguous manner. Specifically, the Fisher information gives the amount of information that an observable random variable provides about an unknown parameter. In our case, the random variable is the position of the particle $\mathbf{x}$, and the unknown parameter is its mass m. In the absence of gravity, observation of the position of the particle can betray information about its mass. For example, a free Gaussian wave packet spreads with variance $\sigma^2(t) = \sigma^2(0) + \hbar t/2m$: one may extract information about its mass by monitoring its position. Formulation of the WEP in terms of Fisher information states that the presence of a gravitational field should not produce more information about the mass of a particle, *i.e.* $F_x(m) = F_x(m)^{\text{free}}$, where $F_x^{\text{free}}(m)$ is the Fisher information in the absence of a gravitational field.

The layout of this chapter is as follows: In Sec. 2, we briefly review the Fisher information and its application to the WEP. In Sec. 3, we look at quantum particles in a Schwarzschild field. In particular, we will investigate the time evolution of Gaussian wavepackets in a static uniform gravitational field, and then generalize this to a static non-uniform gravitational field. We will show how the wavepacket follows the WEP in a static uniform field and its violation in the general case. In Sec. 4, we will derive the time evolution of Gaussian wavepackets in a GW background, and investigate their mass Fisher information. We will show how the wavepacket violates the WEP in the time-dependent uniform field.

2. Fisher Information

In analyzing the statistics of classical experiments that yield random data, one may categorize the variables of the system into *observable* and *latent* variables. Observed variables represent the properties of the system that may be directly measured, and latent variables correspond to the parameters of the mathematical model assumed to describe the stochastic process. Importantly, the latent parameters are by definition intrinsic to the model and cannot be measured, so that for their determination one must resort to the techniques of statistical inference[3] and, in particular, the estimation theory.[4] There is also a third class of variables, which in principle could be measured but are not accessible in practice. These *hidden* variables are important in the context of quantum non-locality, but are beyond the scope of this chapter.

We want to address the WEP notion that the trajectory of a free-falling particle is independent of its mass from an information theory perspective. This can be achieved by observing the positions of the particle, and determining whether extra information about the particle's mass can be extracted in the presence of a gravitational field — the WEP says that no such extra information can be extracted. The formalism that quantifies the amount of information we have about a latent variable in a model, by observing a random variable of that model, is given as the Fisher information.[4] In parameter estimation theory, the Cramer–Rao bound provides the lower bound on the variance of an unbiased estimator $(\tilde{\lambda})$ of the latent variable (λ),

$$\text{var}(\tilde{\lambda}) \geq \frac{1}{nF_x(\lambda)} \tag{3}$$

with n number of observations of random variable x, and the Fisher information defined as

$$F_x(\lambda) = \int dx \ p_\lambda(x) \left[\frac{\partial \log p_\lambda(x)}{\partial \lambda}\right]^2, \tag{4}$$

where $p_\lambda(x)$ is the probability density function for x conditioned on the value of λ. One observes that if p is sharply peaked with respect to changes in λ, it is more probable to identify the true value of λ from x; in other words, x contains a lot of information about λ. This formulation is easily extensible to quantum particles: the position operator leads to a random position variable $\mathbf{x}$, with an associated probability density function $|\psi(\mathbf{x}, t)|^2$. The latent variable that we are interested in is mass m, such that

$$F_x(m) = \int d\mathbf{x} \, |\psi(\mathbf{x}, t)|^2 [\partial_m \log |\psi(\mathbf{x}, t)|^2]^2. \tag{5}$$

We use this mass Fisher information in subsequent sections to quantify the amount of information we can gain about the mass of a quantum particle in various gravitational fields.

3. Quantum Particle in a Schwarzschild Background

In this section, we explore a quantum wavepacket in a gravitational field of Schwarzschild metric. In the first part of the section, we derive the relevant Hamiltonian of the quantum particle, by taking the non-relativistic limit of the Dirac equation under a Schwarzschild metric. In the second part, we consider the case of a uniform gravitational field, relevant for probe quantum particles over short distances and weak gravitational fields. In the third part, we generalize this to non-uniform gravitational fields.

3.1. Non-relativistic limit of the Dirac equation in a Schwarzschild background

We investigate whether extra mass information can be extracted from a Gaussian wavepacket in a Schwarsczhild background. We begin with the Dirac equation in curved space-time, which describes a spin-1/2 particle with rest mass m in a gravitational field,

$$i\hbar\gamma^a e_a{}^\mu(\partial_\mu - \Gamma_\mu)\psi = mc\psi. \tag{6}$$

The space-time metric $g_{\mu\nu}$ can be related at every point to a tangent Minkowski space η_{ab} via tetrads $e^a{}_\mu$, $g_{\mu\nu} = e^a{}_\mu e^b{}_\nu \eta_{ab}$. The tetrads obey the orthogonality conditions $e^a{}_\mu e_a{}^\nu = \delta^\nu_\mu$ and $e^a{}_\mu e_b{}^\mu = \delta^a_b$. We use the convention that Latin indices represent components in the tetrad frame. The spinorial affine connection $\Gamma_\mu = \frac{i}{4}e^a{}_\nu(\partial_\mu e^{\nu b} + \Gamma^\nu{}_{\mu\sigma}e^{\sigma b})\sigma_{ab}$, where $\Gamma^\nu{}_{\mu\sigma}$ is the affine connection and $\sigma_{ab} \equiv \frac{i}{2}[\gamma_a,\gamma_b]$ are the generators of the Lorentz group. γ_a are gamma matrices defining the Clifford algebra $\{\gamma_a,\gamma_b\} = -2\eta_{ab}$, with space-time metric signature $(-,+,+,+)$.

We will consider the Schwarzschild metric in isotopic coordinates ($x^0 \equiv ct$),

$$ds^2 = V^2(dx^0)^2 - W^2(d\mathbf{x}\cdot d\mathbf{x}), \tag{7}$$

where ($r \equiv \sqrt{\mathbf{x}\cdot\mathbf{x}}$) and

$$V = \left(1 - \frac{GM}{2c^2 r}\right)\left(1 - \frac{GM}{2c^2 r}\right)^{-1}, \tag{8}$$

$$W = \left(1 + \frac{GM}{2c^2 r}\right)^2. \tag{9}$$

Under this metric, Obukhov[5] showed that Eq. (6) can be written in the familiar Schrödinger picture $i\hbar\partial_t\psi = H\psi$, with

$$H = \beta mc^2 V + \frac{c}{2}[(\boldsymbol{\alpha}\cdot\mathbf{p})F - F(\boldsymbol{\alpha}\cdot\mathbf{p})], \tag{10}$$

where $\boldsymbol{\alpha} \equiv \gamma^0\boldsymbol{\gamma}, \beta \equiv \gamma^0, F \equiv V/W$, and $\boldsymbol{p} \equiv -i\hbar\nabla$ (∇ is usually the vector differential operator of the three spatial dimensions). Note that M refers to mass that generates the Schwarzshild background, and m is our probe mass.

A means by which to write down the non-relativistic limit of the Dirac Hamiltonian with relativistic correction terms is provided by the Foldy–Wouthuysen (FW) transformation.[6] This is a unitary transformation which separates the upper and lower spinor components. In the FW representation, the Hamiltonian and all operators are block-diagonal (diagonal in two spinors). There are two variants of the FW transformation known as the *standard* FW (SFW)[6] and *exact* FW (EFW)[5, 7–9] transformations. We will use here the EFW transformation, which is efficient and correct at low-orders, and adequate for our purposes. For higher-order corrections one should use the more involved standard FW, as the EFW may produce spurious results at higher- orders, in some instances.[10]

Central to the EFW transformation is the property that when H anti-commutes with $J \equiv i\gamma^5\beta$, $\{H, J\} = 0$, the transformed Hamiltonian is even (even terms do not mix the upper and lower spinor components, odd terms do). Under the unitary transformation $U = U_2 U_1$, where $(\Lambda \equiv H/\sqrt{H^2})$

$$U_1 = \frac{1}{\sqrt{2}}(1 + J\Lambda), \qquad U_2 = \frac{1}{\sqrt{2}}(1 + \beta J), \tag{11}$$

the transformed Hamiltonian is

$$UHU^+ = \frac{1}{2}\beta(\sqrt{H^2} + \beta\sqrt{H^2}\beta) + \frac{1}{2}(\sqrt{H^2} - \beta\sqrt{H^2}\beta)J$$
$$= \{\sqrt{H^2}\}_{\text{even}}\beta + \{\sqrt{H^2}\}_{\text{odd}}J. \tag{12}$$

Observe that as β is an even operator and J is an odd operator, Eq. (12) is an even expression which does not mix the positive and negative energy states. Note that due to non-commutativity of operators, $\sqrt{H^2} \neq H$; in practice $\sqrt{H^2}$ takes a perturbative expansion where the rest mass energy is the dominant term.

Our Hamiltonian satisfies the EFW anti-commutation property. Using the identity $\alpha^i\alpha^j = i\epsilon^{ijk}\sigma_k\mathbf{I}_2 + \delta^{ij}\mathbf{I}_4$, the perturbative expansion of $\sqrt{H^2}$ yields to first order,

$$H \approx mc^2V + \frac{1}{4m}(W^{-1}p^2F + Fp^2W^{-1}). \tag{13}$$

Note that $\sqrt{H^2} = \{\sqrt{H^2}\}_{\text{even}} = H\mathbf{I}_2$ contains only even terms, and therefore $\{\sqrt{H^2}\}_{\text{odd}} = 0$ in Eq. (12).

Taking the weak gravitational field limit so that,

$$V \approx 1 - \frac{GM}{c^2 r}, \quad W \approx 1 + \frac{GM}{c^2 r}, \tag{14}$$

we get ($\mathbf{g} \equiv -GM\mathbf{r}/r^3$)

$$H = mc^2 + \frac{p^2}{2m} + m\mathbf{g} \cdot \mathbf{x}. \tag{15}$$

The Dirac equation in curved space-time will also give rise to a spin-gravity coupling term ($-\frac{\hbar}{2c}\sigma \cdot \mathbf{g}$), which we neglected in Eq. (13). Here we will look at the mass Fisher information of a Gaussian wavepacket in a weak gravitational field, without the higher-order spin-gravity coupling terms, which we can assume to be small.

3.2. *Uniform gravitational field*

The evolution of a quantum particle is governed by the time-evolution operator $U = e^{-iHt/\hbar}$. (For convenience we set $\hbar = 1$ in Secs. 3.2 and 3.3). Taking the Baker–Campbell–Hausdorff expansion of U to second-order, the time-evolution operator in a Schwarzschild background is[1]

$$U \approx \exp\left(\frac{imt^3}{3}\mathbf{g}^2\right) \exp\left(\frac{it^3}{6m}\nabla\mathbf{g} \cdot \nabla\nabla - \frac{\mathbf{g}t^2}{2} \cdot \nabla\right)$$
$$\exp\left(-imt\,\mathbf{g} \cdot \mathbf{x}\right) U_{\text{free}}, \tag{16}$$

where $U_{\text{free}} = \exp(-imc^2 t)\exp(-it\Delta/2m)$ is the free time-evolution operator in the absence of any gravitational field. Note that the $\exp(-imc^2 t)$ term only acts as a constant phase factor in the non-relativistic limit, and therefore can be ignored.[a]

As our gravitational field is spherically symmetric and stationary, we can reduce our problem to one spatial dimension in the radial direction. We consider a Gaussian wavepacket,

$$\psi(\mathbf{x}, 0) = \left(\frac{2}{\pi}\right)^{1/4} e^{-r^2}, \tag{17}$$

---

[a]The notation $\nabla\mathbf{g} \cdot \nabla\nabla$ represents the $(\partial_i g_j)\partial^i\partial^j$, with repeated indices summed from 1 to 3. Similar notations apply to other related expressions in this chapter.

as this is most amenable to comparison with a classical particle. For probe particles traveling over small distances, it is usual to take the terrestrial gravitational field as uniform. We consider the general case in the next section. In the uniform gravitational case,

$$U = \exp\left(\frac{imt^3}{3}\mathbf{g}^2\right)\exp\left(-\frac{\mathbf{g}t^2}{2}\cdot\nabla\right)\exp\left(-imt\,\mathbf{g}\cdot\mathbf{x}\right)U_{\text{free}}. \tag{18}$$

We would like to have the derivative operator acting on the right-hand side of this equation. To achieve this, we make use of Hadamard's lemma,

$$e^A e^B e^{-A} = \exp\sum_{n=0}^{\infty}\frac{\text{ad}_A^n(B)}{n!} \tag{19}$$

where $\text{ad}_A^n(B)$ is the nth order commutator of A with B, that can be used in Eq. (18) to obtain

$$U = \exp\left(\frac{imt^3}{3}\mathbf{g}^2\right)\exp\left(-\frac{\mathbf{g}t^2}{2}\cdot\nabla\right)$$

$$\times \exp\left(-imt\mathbf{g}\cdot\mathbf{x}\right)\exp\left(\frac{\mathbf{g}t^2}{2}\cdot\nabla\right)\exp\left(-\frac{\mathbf{g}t^2}{2}\cdot\nabla\right)U_{\text{free}}, \tag{20}$$

$$= \exp\left(-\frac{imt^3}{6}\mathbf{g}^2\right)\exp\left(-imt\,\mathbf{g}\cdot\mathbf{x}\right)\exp\left(-\frac{\mathbf{g}t^2}{2}\cdot\nabla\right)U_{\text{free}}. \tag{21}$$

Using the fact that the momentum operator is a translation operator in the conjugate position space, the time evolution of the wave function $[\psi(\mathbf{x}, t) = U\psi(\mathbf{x}, 0)]$ is

$$\psi_{\text{g}}(\mathbf{x}, t) = \exp\left(\frac{imt^3}{3}\mathbf{g}^2\right)\exp\left(-imt\,\mathbf{g}\cdot\mathbf{x}\right)\psi_{\text{free}}\left(\mathbf{x} - \frac{\mathbf{g}t^2}{2}, t\right) \tag{22}$$

where $\psi_{\text{free}}(\mathbf{x}, t) = U_{\text{free}}\psi(\mathbf{x}, 0)$ is the free wave function in the absence of a gravitational field. The uniform gravitational field induces a mass-dependent phase factor in Eq. (22). This mass-dependent phase factor however, is not present in the probability distribution, $|\psi_g(\mathbf{x}, t)|^2 = |\psi_{\text{free}}(\mathbf{x} - \mathbf{g}t^2/2, t)|^2$. Therefore, by a change of variable $(\mathbf{u} = \mathbf{x} - \mathbf{g}t^2/2)$, we see that the uniform gravitational field does not produce any extra mass information, i.e.

$$F_x^{\text{g}}(m) = \int d\mathbf{u}\,|\psi_{\text{free}}(\mathbf{u}, t)|^2[\partial_m \log|\psi_{\text{free}}(\mathbf{u}, t)|^2]^2$$

$$= F_x^{\text{free}}(m). \tag{23}$$

The expected position of the wavepacket in a uniform gravitational field is

$$\langle \mathbf{x_g} \rangle = \int_{-\infty}^{\infty} |\psi_g(\mathbf{x}, t)|^2 \mathbf{x} dx = \frac{\mathbf{g}t^2}{2}. \tag{24}$$

This is the geodesic of a freely falling classical particle with no initial momentum in a uniform gravitational field $\mathbf{g}$. As with the classical case, the expected trajectory of the quantum particle is independent of its mass, in alignment with the WEP, and our finding that there is no more mass Fisher information generated in the presence of a static uniform gravitational field.

3.3. *Non-uniform gravitational field*

In the previous section, we considered the case of constant gravitational field, valid for probe particles over small distances and in a weak gravitational field. In this section, we consider the general case of a quantum particle in a Schwarzschild background. The analysis follows the same as the previous section, up until Eq. (16). We would like to have all the derivative operators in this equation acting on the right-hand side. To achieve this, we employ Hadamard's lemma to re-write

$$\exp\left(\frac{-it^3}{6m}\nabla\mathbf{g}\cdot\nabla\nabla - \frac{\mathbf{g}t^2}{2}\cdot\nabla\right)\exp\left(-imt\,\mathbf{g}\cdot\mathbf{x}\right)$$
$$\times \exp\left(\frac{it^3}{6m}\nabla\mathbf{g}\cdot\nabla\nabla + \frac{\mathbf{g}t^2}{2}\cdot\nabla\right) \tag{25}$$

as approximately

$$\exp\left(-\frac{imt^3}{2}\mathbf{g}^2\right)\exp\left(\frac{-imt}{2}\nabla\mathbf{g}^2\cdot\mathbf{g}\right)\exp\left(-imt\mathbf{g}\cdot\mathbf{x}\right)$$
$$\times \exp\left(\frac{t^4}{6}\nabla\mathbf{g}^2\cdot\nabla\right). \tag{26}$$

Using this relation, Eq. (16) becomes

$$U \sim e^{-im\phi(\mathbf{x},t)}e^{\mathcal{D}}U_{\text{free}}, \tag{27}$$

where

$$\phi(\mathbf{x}, t) = \frac{t^3}{6}\mathbf{g}^2 + t\left(\frac{\nabla \mathbf{g}^2 \cdot \mathbf{g}}{2} + \mathbf{g} \cdot \mathbf{x}\right), \tag{28}$$

$$\mathcal{D} = -\frac{it^3}{6m}\nabla \mathbf{g} \cdot \nabla\nabla + \frac{t^4}{6}\nabla \mathbf{g}^2 \cdot \nabla - \frac{\mathbf{g}t^2}{2} \cdot \nabla. \tag{29}$$

Noting that $\psi_{\text{free}}(\mathbf{x}, t) = e^{-\mathcal{D}}(e^{-\mathcal{D}}e^{\mathbf{x} \cdot \nabla}e^{\mathcal{D}})\psi_{\text{free}}(0, t)$, and that the term in the parenthesis can be transformed with Hadamard's lemma, we finally arrive at the time-evolved wavefunction

$$\psi(\mathbf{x}, t) \sim e^{-im\phi(\mathbf{x},t)}\psi_{\text{free}}(\mathbf{x} + \mathbf{d}, t), \tag{30}$$

where $(\mathbf{p} = -i\nabla)$

$$\mathbf{d} = \frac{t^2}{2}(\mathbf{x} \cdot \nabla \mathbf{g} - \mathbf{g}) + \frac{t^3}{3m}\mathbf{p} \cdot \nabla \mathbf{g} + \frac{5t^4}{48}\nabla \mathbf{g}^2. \tag{31}$$

One observes that as displacement $\mathbf{d}$ is a function on mass, the time-evolution of the probability distribution of the wavefunction in the non-uniform gravitational is also dependent on mass. Unlike the previous uniform gravitational field case, the expected position of the wavepacket in a non-uniform gravitational field is no longer independent of mass. Then according to Eq. (5), one can extract extra mass information by observing the particle in a non-uniform gravitational field, violating the WEP.

4. Quantum Particle in Gravitational Waves

In this section, we explore a quantum wavepacket when a gravitational wave (GW) passes through it. In the first part of the section, we derive the relevant Hamiltonian of the quantum particle, by taking the non-relativistic limit of the Dirac equation under a Minkowski metric with perturbations (gravitational waves). In the second part, explore the time evolution of the particle, showing that it violates the WEP.

4.1. Non-relativistic limit of Dirac equation in gravitational waves

The metric for a generally polarized linear plane GW is

$$ds^2 = -c^2 dt^2 + dz^2 + (1 - 2v)dx^2 + (1 + 2v)dy^2 - 2u\,dxdy, \tag{32}$$

where $u = u(t - z)$ and $v = v(t - z)$ are functions which describe a wave propagating in the z-direction. We will consider the case of a circularly polarized GW traveling along the z-direction, i.e. $v = f = f_0 \cos(kz - \omega t)$ and $u = if$. Under this metric, Eq. (6) can be written in the familiar Schrödinger picture as[11, 12]

$$H = \beta mc^2 + c\alpha^j(\delta_j^i + T_j^i)p_i, \tag{33}$$

with

$$T_j^i = \begin{pmatrix} v & -u & 0 \\ -u & -v & 0 \\ 0 & 0 & 0 \end{pmatrix}. \tag{34}$$

Applying an EFW (or SFW) transformation, as in the previous static gravitational field case, and ignoring higher-order terms, one arrives at[2]

$$H_{\text{GW}} = \frac{1}{2m}(\delta^{ij} + 2T^{ij})p_i p_j + mc^2. \tag{35}$$

The Dirac equation in a GW background will also give rise to a spin-gravity coupling term $[\frac{\hbar}{2m}\partial^i(T^{jl})\epsilon_{ijk}\sigma^k p_l]$.[13] As we assume that the gravitational wavelength is much larger than the atomic dimensions, $\partial^i(T^{jl})$ is small.

4.2. Time-dependent gravitational field

We would like to know how a Gaussian wavepacket behaves in a GW background. We will consider the wavepacket located at $z = 0$, in one spatial dimension x, without loss of generality in our conclusions,[b]

$$\psi(x, 0) = \left(\frac{2}{\pi}\right)^{1/4} e^{-(x-x_0)^2}. \tag{36}$$

We apply the unitary transformation operator $U = \exp[-i/\hbar \int H_{\text{GW}}(t)dt]$ to Eq. (36) to get the time evolution of a wave packet in a GW background. Specifically, we Fourier transform the wavepacket into momentum space, $\psi(k, 0) = \left(\frac{1}{2\pi}\right)^{1/4} e^{-k^2/4} e^{-ikx_0}$. This allows us to easily

---

[b]One notes that cigar shape potential traps could confine a particle to an effective one spatial dimension.

write down the time evolution of the wavepacket in momentum space,

$$\psi_{\rm GW}(k,t) = e^{-i/\hbar \int_o^t H_{\rm GW}(t')dt'} \psi(k,0)$$

$$= \left(\frac{1}{2\pi}\right)^{1/4} e^{-i\hbar k^2 [t+f_0 \sin(\omega t)/\omega]/2m} e^{-k^2/4} e^{-ikx_0}. \tag{37}$$

where we have used the fact that $H_{\rm GW}$ in one dimension is $H_{\rm GW} = \frac{1+f_0 \cos \omega t}{2m} p^2$. We Fourier transform back into position space to arrive at

$$\psi_{\rm GW}(x,t) = \left(\frac{2}{\pi}\right)^{1/4} \frac{e^{-(x-x_0)^2/b}}{\sqrt{b}}, \tag{38}$$

where we have defined

$$b \equiv 1 + \frac{2i\hbar}{m}[t + f_0 \sin(\omega t)/\omega]. \tag{39}$$

From Eq. (38), the expected position of the wavepacket in a GW background is

$$\langle x_{\rm GW} \rangle = x_0. \tag{40}$$

In other words, the particle is expected to remain at rest in the co-ordinate system of Eq. (32). This actually is not surprising as this is also what happens in the classical case. In this case, the presence of a GW is measured in the change of the proper distance between two particles. However, unlike the classical case, the presence of the GW will generate mass information. From Eq. (38), we numerically calculate the mass Fisher information (Eq. (5)) of the particle in the GW, and compare it to the free case. To reveal the effects of the GW, f_0 and m are set to unity (and $\hbar = 1$). More realistic values are discussed later. Figure 1 plots the difference in the mass Fisher information in a GW background from the free case, showing that in general it is different from zero. This means that one can extract mass information of the particle from the GW, in violation of the WEP, in stark contrast to the static uniform gravitational field of the previous section, i.e.

$$F_x^{\rm GW}(m) \neq F_x^{\rm free}(m). \tag{41}$$

We have thus shown that quantum particles violate the WEP in principle. In the next section, we discuss how such violations may be practically exploited.

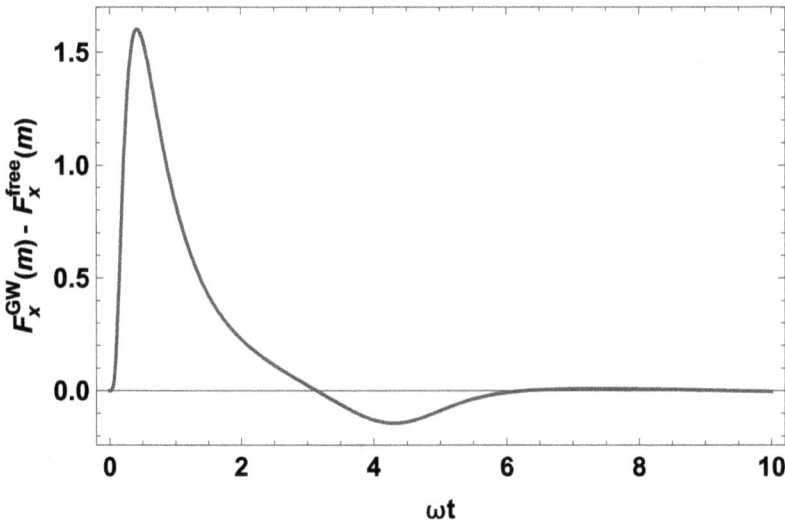

Fig. 1.   The difference in the mass Fisher information in a GW background from the free case, over time. As the difference can be non-zero, one can extract mass information of the particle from the GW, in violation of the WEP.

5. Experiments

This section explores how WEP violations in quantum systems may be used in experiments. Specifically, we discuss the potential of such violation as the basis for GW detection, and in the observation of the gravitational Casimir effect.

5.1. *Gravitational wave detector*

As a GW can betray the mass of a quantum particle, it is tantalizing to ask whether one could use this fact to detect the presence of a GW. As a simple case study, let us consider a particle detector that is at rest in the local co-ordinate system, measuring the probability of a particle being at position x_0. At time t_0, we place a particle at x_0. In the absence of the GW, the probability of detecting the particle always decreases as the wave function spreads according to Schrdinger's equation, as shown by the dotted line in Fig. 2. In the presence of the GW, the behavior of the particle is distinctly different; in particular the probability of detecting the particle at x_0 can increase, as shown by the solid line in Fig. 2. Therefore, in principle one could simply look for these characteristic increases in the probability of particle detection as signatures of a GW. In practice, however,

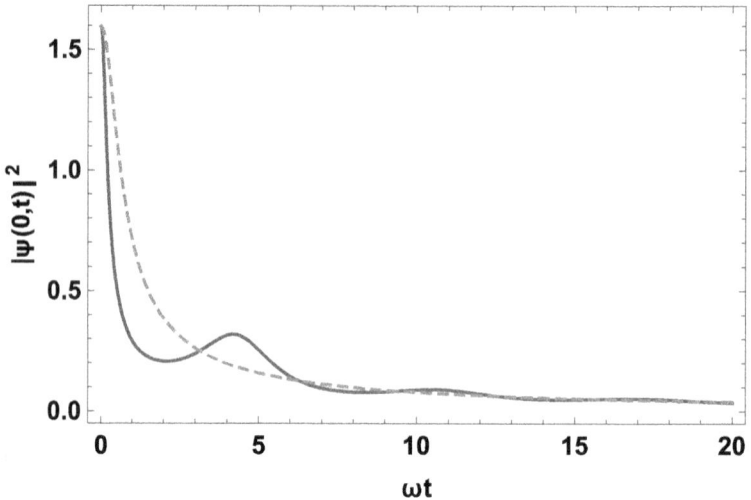

Fig. 2. The probability of detecting a particle (at $x = 0$) always decreases over time for a free particle in the absence of a GW (dotted line). In comparison, the probability of detecting the same particle can increase in the presence of the GW (solid line).

these characteristic increases in the probability of detection are restrictively small for known GW sources. For example, let us consider a rubidium atom ($m = 1.4 \times 10^{-25}$ kg) in a GW generated by the PSR B1913+16 binary pulse system, which has an amplitude of $f_0 = 10^{-14}$ on Earth and the frequency $\omega = 2.4 \times 10^{-4}$.[13] To detect this GW, one would require a probability detection resolution of 10^{-29}. With such a small change in the probability distribution, it is unlikely that observing a *single* quantum particle in this manner can betray the presence of a GW. This however, does not preclude more sophisticated setups in utilizing quantum WEP violation for GW detection, such as ensembles of interacting particles,[14] high precision measurements with quantum metrology,[15] or other novel setups: of note are works that propose that WEP violating Cooper pairs in superconductors may give rise to GW detectors,[16–23] transducers,[24,25] and mirrors.[26–28] In these works, it is conjectured that the delocalized Cooper pairs behave differently from the localized ion cores of the superconductor in the presence of a GW, due to the larger degree of WEP violation in the former. We discuss this in more detail in the next section.

5.2. *Gravitational Casimir effect*

There have been suggestions that the properties of *quantum fluids* (superconductors, superfluids, quantum Hall fluids, Bose–Einstein condensates) may enhance the interaction with GW, leading to superfluids as a medium for gravitational antennae,[16–20, 22, 23] superconducting circuits as GW detectors,[21] transducers,[24, 25] and mirrors.[26–28] These ideas have not been met without controversy.[29–31] The reason for this is that many of these ideas heuristically apply the notion that quantum particles violate the WEP. Here we provide a more rigorous characterization of the WEP for quantum particles in GWs.

We can define a parameter β^A to measure the magnitude of violation of WEP for a quantum system A in GWs with mass Fisher information $F_x^A(m)]$ as[32]

$$\beta^A = \frac{\int |F_x^A(m) - F_x^{\text{free}}(m)| dt}{\int |F_x^{\text{free}}(m)| dt}. \tag{42}$$

Now for two quantum systems A and B, if $\beta^A > \beta^B$, then the magnitude of violation of WEP is larger in A than B. This violation of the WEP may be experimentally detected in superconductors because the delocalization of Cooper pairs (CPs) are expected to violate the WEP more than localized particles. This can be observed by considering the two-particle wavefunction of the CP[33]

$$\psi_0(\mathbf{r}^{(1)}, \mathbf{r}^{(2)}) = \sum_{\mathbf{k}} g_{\mathbf{k}} e^{i\mathbf{k}\cdot\mathbf{r}^{(1)}} e^{-i\mathbf{k}\cdot\mathbf{r}^{(2)}}, \tag{43}$$

where $\mathbf{r}^{(i)}$ is the coordinate of particle i. We can now use the unitary transformation operator $U = \exp(-iH_{\text{GW}}(\mathbf{p}^{(1)}, \mathbf{p}^{(2)})t)$ to get the time evolution of the CP in a GW background, where

$$H_{\text{GW}}(\mathbf{p}^{(1)}, \mathbf{p}^{(2)}) = \frac{1}{2m_{\text{CP}}}(\delta^{ij} + 2T^{ij})(p_i^{(1)} p_j^{(1)} + p_i^{(2)} p_j^{(2)}) + m_{\text{cp}}c^2 + V, \tag{44}$$

with V as the interaction with the ionic lattice. As the CPs are delocalized relative to the localized ionic lattice (shown in Fig. 3(a)), they exhibit a larger magnitude of WEP violation, $\beta^{\text{Cooper}} > \beta^{\text{ion}}$, because

$$\beta^{\text{Cooper}} = \frac{\int |F_x^{\text{Cooper}}(m_{\text{cp}}) - F_x^{\text{free}}(m_{\text{cp}})| dt}{\int |F_x^{\text{free}}(m_{\text{cp}})| dt}$$

$$> \frac{\int |F_x^{\text{ion}}(m_{\text{ion}}) dt - F_x^{\text{free}}(m_{\text{ion}})| dt}{\int |F_x^{\text{free}}(m_{\text{ion}})| dt} = \beta^{\text{ion}}. \tag{45}$$

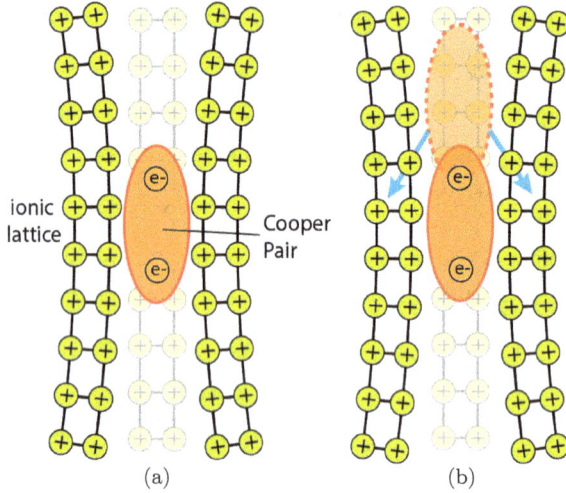

Fig. 3. Mechanism for spectral reflection of GWs from superconductors. (a) The negatively-charged CP deforms the positively-charged ionic lattice. (b) The GW accelerates the delocalized CP relative to the lattice. However, the positively-charged ionic lattice suppresses this acceleration, thereby partially reflecting the GW.

Thus, GWs will tend to accelerate the CPs relative to the ionic lattice due to the larger magnitude of WEP violation in CPs relative to the ionic lattice, as depicted in Fig. 3(b). As the CPs and ionic lattice are oppositely charged, they will resist any charge separation, thereby (partially) reflecting the GW, analogous to the way conductors repel electromagnetic waves. It was initially proposed that such a reflection of GWs occurs due to CPs not moving on the geodesics on which ions move;[26, 34–36] such a conjecture may be quantified as the difference in the mass Fisher information of the CPs and the ion core.

GWs with wavelengths of the same order as the CP coherence length would be required to test this WEP violation, but such GWs are difficult to generate in controlled experiments. However, one can utilize the full spectrum of the virtual GWs (formed from virtual gravitons in the vacuum) in a gravitational Casimir effect to test such violations of WEP.[27, 37] This can be explicitly demonstrated using the Einstein field equations linearized around a flat space-time metric (which resembles the Maxwell equations with a magnetic monopole),[38–42]

$$\nabla \cdot \mathbf{E} = \kappa \rho^{(E)}, \quad \nabla \cdot \mathbf{B} = \kappa \rho^{(M)}, \tag{46}$$

$$\nabla \times \mathbf{E} = -\frac{\partial \mathbf{B}}{\partial t} - \kappa \mathbf{J}^{(M)}, \quad \nabla \times \mathbf{B} = \frac{\partial \mathbf{E}}{\partial t} + \kappa \mathbf{J}^{(E)}, \tag{47}$$

where $\kappa \equiv 8\pi G/c^4$, $E_{ij} \equiv C_{0i0j}$, $B_{ij} \equiv \star C_{0i0j}$, $C_{\alpha\beta\mu\nu}$ is the Weyl tensor, $\star$ denotes Hodge dualization[c], and $\rho_i^{(E)} \equiv -J_{i00}$, $\rho_i^{(M)} \equiv -\star J_{i00}$, $J_{ij}^{(E)} \equiv J_{i0j}$, $J_{ij}^{(M)} \equiv \star J_{i0j}$[d].

For parallel plates separated by a distance b along the z-axis (with $k_\parallel \equiv \sqrt{k_x^2 + k_y^2}$), the gravitational Casimir energy can be obtained by summing over $(\omega_n^+, \omega_n^\times)$, which are the relevant modes of virtual GWs for the system,[37,43]

$$E_0(b) = \frac{\hbar}{4\pi} \int_0^\infty k_\parallel dk_\parallel \sum_n (\omega_n^+ + \omega_n^\times)\sigma, \tag{48}$$

where σ is the surface area of the plates. One notes that the calculated energy diverges due to the summation over the infinite number of allowed modes, and therefore one needs to renormalize it by subtracting the vacuum energy at infinite separation[27]

$$E_R(b) = \frac{E_0}{\sigma} - \lim_{b \to \infty} \frac{E_0}{\sigma}. \tag{49}$$

It is this finite renormalized gravitational Casimir energy that can be detected using superconducting parallel plates.[44] The strength of the gravitational Casimir force will depend on the difference between the magnitude of WEP violation experienced by the delocalized CP (β^{Copper}) and the localized ionic lattice (β^{ion}). Although the gravitational Casimir effect is expected to be smaller than the electromagnetic Casimir effect, it can be measured from the additional force produced at the onset of superconductivity (which is experimentally easier than measuring absolute Casimir forces). In fact, the magnitude of this gravitational Casimir effect has been estimated to be on the same order as the electromagnetic Casimir effect using heuristic calculations.[26,27,34–37] However, as it is important to determine the exact magnitude of this gravitational Casimir effect, one would need to rigorously perform these calculations; this has not yet been done. It may be noted that even if the exact magnitude of the gravitational Casimir

---

[c]Dualization is defined by $\star C_{\mu\nu\rho\sigma} \equiv \frac{1}{2}\epsilon_{\mu\nu\alpha\beta}C_{\rho\sigma}{}^{\alpha\beta}$, $\star J_{\mu\nu\rho} \equiv \frac{1}{2}\epsilon_{\mu\nu\alpha\beta}J^{\alpha\beta}{}_\rho$, with $\epsilon_{\mu\nu\alpha\beta}$ as the Levi-Civita tensor.

[d]The matter current $J_{\mu\nu\rho} \equiv (\eta_{\rho[\mu}T_{,\nu]}/3) - T_{\rho[\mu,\nu]}$ is obtained from stress perturbations $T_{\mu\nu}$ ($T \equiv T_\mu^\mu$).

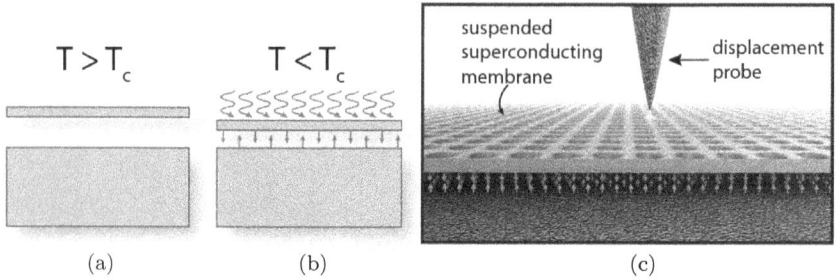

Fig. 4. General schematic representations of experiments: (a) A superconducting nanomembrane is fabricated above a superconducting substrate. (b) Once cooled below the superconducting transition temperature T_c, one would observe the sudden change in distance between membrane and substrate due to the gravitational Casimir effect. (c) One could measure this change in displacement using a number of currently available displacement probes, which can measure distances on the atomic scale.

effect is reduced by several orders of magnitude, from what has been calculated using heuristic calculations, it should still be measurable.

A gravitational Casimir experiment with superconductors would be relatively simple, as shown in Fig. 4. On a microchip, one can achieve remarkable parallelism,[44] making it possible to fabricate a free-standing superconducting nanomembrane over a superconducting substrate.[45] The entire microchip can then be cooled below the superconducting transition temperature T_c, experiencing an additional gravitational Casimir force, as depicted in Fig. 4(a, b). Any small additional forces at T_c would lead to large displacements of these ultra-sensitive nanomembranes due to their high-aspect ratio.[45] A displacement probe such as those used in scanning tunneling microscopy,[46] or a photonic crystal probe,[47] could measure displacements of this membrane with atomic precision, as depicted in Fig. 4(c). The magnitude of WEP violation in CPs depends on the CP coherence length, which in turn depends on the properties of superconductors,[48,49] so one may use various superconducting materials to measure the gravitational Casimir effect with current technology. As the WEP is a building block of general relativity, convincingly demonstrating its violation would signify a drastic departure from our current understanding of gravity at short distances.

References

1. L. Seveso and M. G. A. Paris, *Ann. Phys.* **380**, 213 (2017).
2. J. Q. Quach, *Eur. Phys. J. C* **80**, 10 (2020).

3. L. Wasserman, *All of Statistics: A Concise Course in Statistical Inference.* Springer Texts in Statistics. Springer New York, 2013.
4. S. M. Kay, *Fundamentals of Statistical Signal Processing: Detection Theory.* Fundamentals of Statistical Si. Prentice-Hall PTR, 1998.
5. Y. N. Obukhov, *Phys. Rev. Lett.* **86**, 192 (2001).
6. L. L. Foldy and S. A. Wouthuysen, *Phys. Rev.* **78**, 29 (1950).
7. E. Eriksen, *Phys. Rev.* **111**, 1011–1016, Aug. 1958.
8. A. G. Nikitin, *J. Phys. A: Math. Nucl. Gen.* **31**, 3297 (1998).
9. U. D. Jentschura and J. H. Noble, *J. Phys. A: Math. Theor.* **47**, 045402 (2014).
10. U. D. Jentschura and J. H. Noble, *Phys. Rev. A* **88**, 022121 (2013).
11. B. Gonçalves, Y. N. Obukhov and I. L. Shapiro, *Phys. Rev. D* **75**, 124023 (2007).
12. J. Q. Quach, *Phys. Rev. D* **92**, 084047 (2015).
13. J. Q. Quach, *Phys. Rev. D* **93**, 104048 (2016).
14. M. Napolitano, M. Koschorreck, B. Dubost, N. Behbood, R. J. Sewell and M. W. Mitchell, *Nature* **471**, 486 (2011).
15. V. Giovannetti, S. Lloyd and L. Maccone, *Phys. Rev. Lett.* **96**, 010401 (2006).
16. G. Papini, *Lettere al Nuovo Cimento* **4**, 1027 (1970).
17. J. Anandan, *Phys. Rev. Lett.* **47**, 463 (1981).
18. R. Y. Chiao, *Phys. Rev. B* **25**, 1655 (1982).
19. J. Anandan and R. Y. Chiao, *Gen. Relativ. Gravit.* **14**, 515 (1982).
20. J. Anandan, *Phys. Rev. Lett.* **52**, 401 (1984).
21. J. Anandan, *Phys. Lett. A* **110**, 446 (1985).
22. H. Peng, *Gen. Relativ. Gravit.* **22**, 609 (1990).
23. H. Peng, D. G. Torr, E. K. Hu and B. Peng, *Phys. Rev. B* **43**, 2700 (1991).
24. R. Y. Chiao. Conceptual tensions between quantum mechanics and general relativity: Are there experimental consequences. In J. D. Barrow, P. C. W Davies and C. L. Harper Jr, editors, *Science and Ultimate Reality: Quantum Theory, Cosmology, and Complexity*, 254–279. Cambridge University Press, Cambridge (2004).
25. A. L. Licht, *gr-qc/0402109*.
26. S. J. Minter, K. Wegter-McNelly and R. Y. Chiao, *Physica E* **42**, 234 (2010).
27. J. Q. Quach, *Phys. Rev. Lett.* **114**, 081104 (2015).
28. J. Q. Quach, *Phys. Rev. Lett.* **118**, 139901 (2017).
29. M. Kowitt, *Phys. Rev. B* **49**, 704 (1994).
30. E. G. Harris, *Found. Phys. Lett.* **12**, 201 (1999).
31. C. Kiefer and C. Weber, *Ann. Phys.* **14**, 253 (2005).
32. S. Bahamonde, M. Faizal, J. Q. Quach and R. A. Norte, *Int. J. Mod. Phys. D* **29**, 2043024 (2020).
33. M. Tinkham, *Introduction to Superconductivity.* Courier Corporation, 2004.
34. N. A. Inan, J. J. Thompson and R. Y. Chiao, *Fortsch. Phys.* **65**, 1600066 (2017).
35. H. Yu, Z. Yang and P. Wu, *Phys. Rev. D* **97**, 026008 (2018).
36. N. A. Inan, *Int. J. Mod. Phys. D* **26**, 1743031 (2017).
37. J. Hu and H. Yu, *Phys. Lett. B* **767**, 16 (2017).

38. A. Matte, *Can. J. Math.* **5**, 1 (1953).
39. W. B. Campbell and T. Morgan, *Physica* **53**, 264 (1971).
40. R. Maartens and B. A. Bassett, *Class. Quant. Grav.* **15**, 705 (1998).
41. J. Ramos, M. Montigny and F. C. Khanna, *Gen. Rel. Grav.* **42**, 2403 (2010).
42. P. Szekeres, *Ann. Phys.* **64**, 599 (1971).
43. M. Bordag, G. L. Klimchitskaya, U. Mohideen and V. M. Mostepanenko, *Int. Ser. Monogr. Phys.* **145**, 1 (2009).
44. R. A. Norte, M. Forsch, A. Wallucks, I. Marinković and S. Gröblacher, *Phys. Rev. Lett.* **121**, 030405 (2018).
45. J. P. Moura, R. A. Norte, J. Guo, Clemens Schäfermeier and S. Gröblacher, *Opt. Express* **26**, 1895 (2018).
46. I. Battisti, G. Verdoes, K. v. Oosten, K. M. Bastiaans and M. P Allan, *Rev. Sci. Instrum.* **89**, 123705 (2018).
47. L. Magrini, R. A. Norte, R. Riedinger, I. Marinković, D. Grass, U. Delić, S. Gröblacher, S. Hong and M. Aspelmeyer, *Optica* **5**, 1597 (2018).
48. J. J. Lee, F. T. Schmitt, R. G. Moore, S. Johnston, Y.-T. Cui, W. Li, M. Yi, Z. K. Liu, M. Hashimoto, Y. Zhang, *et al.*, *Nature* **515**, 245 (2014).
49. J. G. Bednorz and K. A. Müller, *Z. Phys. B Condens. Matter* **64**, 18 (1986).

https://doi.org/10.1142/9789811253591_0005

Chapter 5

A Critical Discussion on Equivalence Principle in Theories of Gravity

Salvatore Capozziello[*,†,‡,‖] and Gaetano Lambiase[§,¶,**]

*Dipartimento di Fisica "E. Pancini", Universita' di Napoli
"Federico II", Via Cinthia I-80126 Napoli, Italy
†Scuola Superiore Meridionale, Largo S. Marcellino 10,
I-80138 Napoli, Italy
‡INFN, Sezione di Napoli, C.U. Monte S. Angelo, Via Cinthia,
I-80126 Napoli, Italy
§Dipartimento di Fisica E.R. Cainaiello, Universita' di Salerno,
Via Giovanni Paolo II, I 84084-Fisciano (SA), Italy
¶INFN, Gruppo Collegato di Salerno, Sezione di Napoli,
Via Giovanni Paolo II, I 84084-Fisciano (SA), Italy
‖capozziello@na.infn.it
**lambiase@sa.infn.it

Physical laws, ranging from cosmology to local scales, cannot be properly formulated without taking into account concepts and formalism of General Relativity. The latter is based on the Equivalence Principle representing one of the most fundamental principles of Nature. Validity or violation of the Equivalence Principle represents one of the main challenges of modern physics both from theoretical and experimental points of view. In this respect, high precision experiments are conceived and realized for testing Einstein's theory as well as its generalizations by using established and novel methods, such as, for example, experiments based on a large variety of sensors like atomic clocks, accelerometers, gyroscopes, gravimeters, etc.

Here, we discuss the foundation of the Einstein Equivalence Principle, both in the weak and strong formulations. The discussion is enlarged to the case of extended and alternative theories of gravity. Moreover, we present the possibility of violation of the Equivalence Principle in quantum field theories at finite temperature, both in the frameworks of General Relativity and modified gravity. A possible test of the Einstein equivalence principle obtained observing the photon ring of a black hole is discussed.

1. Introduction

The basic idea of General Relativity (GR) is that space and time are entangled into a unique structure, i.e. the space-time, with a pseudo-Riemannian manifold structure endowed with a Lorenzian signature, and such that the flat Minkowski space-time is recovered in absence of gravitational fields. Moreover, since GR is an extension of classical mechanics, it has to reproduce, in the weak-field regime, results of Newton's physics (i.e. the dynamics related to the planetary motion and the self-gravitating structures such as stars, galaxies, clusters of galaxies), and pass observational tests in the Solar System (the so-called *classical tests* of GR).[1,2] As a fundamental theory of gravity, GR is required to explain some issues like: (1) Galactic dynamics, considering baryonic constituents, like stars, planets, dust and gas, radiation; (2) the large scale structure formation; (3) dynamics of the whole Universe reproducing cosmological parameters like the Hubble expansion rate, the density parameters and the behavior of cosmic fluid.

The Galileo experiments on the free-fall of different bodies lead to the conclusion that gravitational and inertial mass ratio are identical for different bodies (the acceleration of free-falling massive bodies is independent of their nature). This result is crucial for Einstein's GR as well as for any metric theory of gravity. To summarize, GR is based on four main assumptions[1,2]:

- The *"Relativity Principle"* — there is no preferred frames in Physics, i.e. all frames (accelerated or not) are equivalent.
- The *"Equivalence Principle"* (EP) — inertial effects are locally indistinguishable from gravitational effects, which means the equivalence between inertial and gravitational masses and, furthermore, any gravitational field, can be locally canceled finding out a local frame where accelerations are null.
- The *"General Covariance Principle"* — field equations must be "covariant" in form, i.e. they must be invariant under the action of any space-time diffeomorphism.
- The *"Causality Principle"* — each point of space-time admits a universal notion of past, present, and future.

Based on these assumptions, Einstein postulated that the gravitational field can be described, in a four-dimensional space-time, by the metric tensor $g_{\mu\nu}$ entering the line element $ds^2 = g_{\mu\nu}dx^\mu dx^\nu$, with the same signature of Minkowski metric. The metric components $g_{\mu\nu}$ are related to the gravitational potentials, while space-time is curved owing to the distribution of energy–matter sources.

The above principles require that the space-time structure is determined by a Lorentzian metric g or by the couple g and a connection Γ (Einstein assumed a torsionless manifold because, at that time, the spin of particles was not considered a possible source for the gravitational field). The metric g fixes the space-time causal structure, that is the light cones. The connection Γ fixes instead the free-fall of objects (that is the local inertial observers according to the Equivalence Principle and then the geodesic structure). In general, Γ and g can be independent.[3] However, in GR Γ is the Levi-Civita connection of g, and therefore they are related by some compatibility relations.

Here some comments are in order. In formulating GR, the metric g describes gravity (it is the fundamental object giving rise to dynamics). The connection $\Gamma^\alpha_{\mu\nu} = \left\{ {\alpha \atop \mu\nu} \right\}_g$ has no dynamics. The metric g determines the causal structure (light cones), the measurements (rods and clocks) and the free-fall of test particles (geodesic structure). In such a case, space-time is determined by $\{\mathcal{M}, g\}$, that is constituted by a (pseudo-)Riemann manifold and a metric. According to the EP, the metric structure is associated to the connection Γ since it can be switched off and set to zero at least in a point (local Minkowskian space-time). On the other hand, the Palatini formalism assumes a (symmetric) connection Γ and a metric g which can vary independently. This picture implies that space-time is a triple $\{\mathcal{M}, g, \Gamma\}$ where the metric g determines causal structure and Γ, the free-fall. In the Palatini formalism, connections are differential equations determining dynamics. Therefore, the connection is the fundamental field in the Lagrangian while the metric g enters the Lagrangian in view to define lengths and distances, so that it defines the causal structure but has no dynamical role. As a consequence, there is no reason to assume g as the potential of Γ as in the Levi-Civita case.

Despite the self-consistency and the solid foundation, as well as all confirmed predictions of GR, there are several issues, both from theoretical and experimental viewpoints, which make GR incomplete. GR, in fact, is not capable of addressing completely Galactic, extra-galactic, and cosmic dynamics because some exotic forms of matter–energy, generically called

Dark Matter (DM) and *Dark Energy* (DE), must be invoked. DM and DE constitute up to 95% of the total amount of cosmic matter–energy but, at the moment, also if their effects are evident at large scales, there is no final evidence of their existence at the fundamental level.[3,4] From this perspective, instead of changing the energy-matter sector of the Einstein field equations, one can ask for modifications of the geometrical sector. In this view, GR can be extended by including geometric invariants derived, e.g. from curvature and torsion, into the Hilbert–Einstein Action. Such additional invariants occur, for example, at the fundamental level considering quantum fields in curved space-time.[3,5] At the moment, however, both dark matter–energy picture and extended (alternative) gravity[a] are valid candidates to describe the observed anomalies that GR cannot explain.

With this consideration in mind, we discuss here the role of EP in the debate of theories of gravity both from theoretical and experimental points of view.

This chapter is organized as follows. In Sec. 2, we review different formulations of EP. After summarizing the main topics of GR and Quantum Field Theory (QFT) in curved space-times, we discuss metric theories of gravity considering possible extensions and modifications of GR. Motivations, both theoretical and experimental, suggesting generalizations of GR, are considered. These theories have been introduced to account for shortcomings of GR at cosmological and astrophysical scales addressed by Inflation, Dark Matter, and Dark Energy. Moreover, GR does not include quantum effects. It is then natural to ask whether the EP still holds whether quantum physics is taken into account. In this context, we review the EP by considering QFT at finite temperature, both in GR and in some extended gravity (we shall refer in particular to Brans–Dicke theory). In Section 4, we discuss the Strong Equivalence Principle. Section 5 is devoted to the test Einstein Equivalence Principle (EEP) via recent Event Horizon Telescope (ETH) data. Conclusions are drawn in Sec. 6.

---

[a] An important remark is useful at this point. With the term *Extended Gravity*, we mean any class of theories by which it is possible to recover Einstein GR as a particular case or in some post-Einstenian limit as in the case of $f(R)$ gravity. With *Alternative Gravity*, we mean a class of theories which considers different approaches with respect to GR, for example, the Teleparallel Equivalent Gravity that considers the torsion scalar instead of curvature scalar to describe dynamics.

2. The Foundation of Equivalence Principle

The EP plays a relevant role to discriminate among concurring theories of gravity. The roles of $\{g, \Gamma\}$ are related to the validity of EP. Precise measurements of EP could provide information on the nature of Γ, that is if it is Levi-Civita or if it is a more general connection, not related to g. Furthermore, possible violations of EP allow to put in evidence other dynamical fields, such as torsion, and discriminate among the fundamental structures of space-time (Riemannian structure, as in GR, or not).

The relevance of EP comes from the following points:

- The EP could discriminate among competing theories of gravity;
- EP holds at the classical level, but it could be violated at the quantum level;
- EP allows to investigate independently geodesic and causal structure of space-time. If it is violated at the fundamental level, such structures could be independent.

From a theoretical point of view, EP constitutes the foundation of metric theories. The first formulation of EP comes out from the formulation of gravity by Galileo and Newton, which states that the inertial mass m_i and the gravitational mass m_g of physical objects are equivalent. From this statement, one can derive the Weak Equivalence Principle (WEP) which implies that it is impossible to distinguish, locally, between the effects of a gravitational field from those experienced in uniformly accelerated frames using the straightforward observation of the free-fall of physical objects.

A first generalization of WEP states that Special Relativity is locally valid. Einstein obtained, in the framework of Special Relativity, that mass can be reduced to a manifestation of energy and momentum. As a consequence, it is impossible to distinguish between a uniform acceleration and an external gravitational field, not only for free-falling objects, but whatever is the experiment. According to this observation, Einstein EP states:

- The WEP is valid.
- The outcome of any local non-gravitational test experiment is independent of the velocity of free-falling apparatus.
- The outcome of any local non-gravitational test experiment is independent of where and when it is performed in the Universe.

One can define a "local non-gravitational experiment" as that performed in a small-size of a free-falling laboratory. Immediately, it is possible to realize that gravitational interaction depends on the curvature of space-time. It means that the postulates of metric gravity theories have to be satisfied. Hence, the following statements hold:

- Spacetime is endowed with a metric $g_{\mu\nu}$ whose entries constitute the dynamic variables (gravitational potentials).
- The world lines of test bodies are geodesics of the metric.
- In local freely falling frames, i.e. the local Lorentz frames, the non-gravitational laws of physics are those of Special Relativity.

One of the predictions of this principle is the gravitational red-shift, experimentally probed by Pound and Rebka.[1] It is worth noticing that gravitational interactions are excluded from WEP and Einstein EP.

To classify extended and alternative theories of gravity, the gravitational WEP and the Strong Equivalence Principle (SEP) are introduced. The SEP extends the Einstein EP by including all the laws of physics. It states:

- WEP is valid for self-gravitating bodies as well as for test bodies (gravitational WEP).
- The outcome of any local test is independent of the velocity of the free-falling apparatus.
- The outcome of any local test is independent of where and when it is performed in the Universe.

The Einstein EP is recovered from SEP as soon as the gravitational forces are neglected. Several authors claim that the only theory coherent with SEP is GR and then WEP has to be deeply investigated.

A very important issue is the consistency of EP with respect to Quantum Mechanics. GR is not the only gravity theory and several alternatives have been investigated starting from the 1960s.[3] Some of them are effective descriptions coming from quantum field theories on curved space-time.[5]

Considering the space-time as specially relativistic at a background level, gravitation can be treated as a Lorentz-invariant perturbation field on the background. Assuming the possibility of GR extensions and alternatives, two different classes of experiments can be conceived:

- Tests for the foundations of gravity according to the various formulations of EP.
- Tests of metric theories where space-time is endowed with a metric tensor and where the Einstein EP is assumed valid.

The difference between the two classes of experiments consists in the fact that EP can be postulated "a priori" or "recovered" from the self-consistency of the theory. What is clear is that, for several fundamental reasons, extra fields are necessary to describe gravity with respect to other interactions. Such fields can be scalar fields or higher-order corrections of curvature and torsion invariants.[3] For these reasons, two sets of field equations can be considered: The first couples the gravitational field to non-gravitational fields, i.e. the matter distribution, the electromagnetic fields, etc. The second set of equations considers dynamics of non-gravitational fields. In the framework of metric theories, these laws depend only on the metric and this is a consequence of the Einstein EP.

In the case where gravitational field equations are modified with respect to the Einstein ones, and matter fields are minimally coupled with gravity, we are dealing with the *Jordan frame*. In the case where Einstein field equations are preserved and matter fields are non-minimally coupled, we are dealing with the *Einstein frame*. Both frames are conformally related but the very final issue is to understand if passing from one frame to the other (and vice versa) is physically significant. Clearly, EP plays a fundamental role in this discussion. In particular, the main question is if EP is valid in any case or it is violated at the quantum level.

2.1. *The debate on gravitational theories*

As pointed out before, GR provides a unified description of space, time, gravity, and matter. It is formulated in such a way that space and time are entangled quantities which yield the conception of the Universe as a dynamical system.

GR is considered the best theory for describing the gravitational interaction. It has been tested at different scales, ranging from Solar System[6–8] to astrophysics and cosmological scales. More recently, observation has led to two important predictions of GR, that are the Gravitational Waves (GWs) by LIGO/VIRGO Collaboration, and the shadow of Black Hole (BH) confirmed by the Event Horizon Telescope (EHT).[9–11] Despite these successes, GR is not free of shortcomings arising at cosmological/astrophysical scales (infrared scales) and at ultraviolet scales. In the first case, the Big Bang singularity, the flatness, horizon, and monopole problems[12] provided a clear indication that the standard cosmological model, that is the cosmology based on GR and the Standard Model of particles, is inadequate to describe the Universe in extreme high energy-curvature regimes. On the other hand,

GR does not work as a fundamental theory of gravity if the quantum principles are included since Einstein's theory is a classical theory. These reasons, and, the lack of a self-consistent quantum theory of gravity, led to the formulation of various alternatives and extensions of GR. The idea is to formulate an effective theory where GR and its positive results are recovered in some limit. In this respect, *Extended Theories of Gravity* (ETGs) have become a paradigm to study the gravitational interaction, in which higher order curvature invariants and minimally or non-minimally coupled scalar fields are taken into account into dynamics.[3, 13]

Other reasons to modify GR are related to the issue of incorporating Mach's principle into the theory. Mach's principle states that local inertial frames are determined by the average motions of distant astronomical objects.[14] The gravitational coupling, therefore, is influenced by the surrounding matter distribution and becomes a function of space-time through the introduction of a scalar field. As a consequence, the concept of inertia and the formulation of the EP need to be accordingly revised. The first example of alternative gravity is provided by the Brans–Dicke theory.[15] In this theory, the Mach principle is naturally incorporated. The gravitational constant turns out to be dependent on a scalar field non-minimally coupled to the geometry (it is assumed "variable").[15-17] Non-minimal couplings to the geometry, as effective actions, occur from any scheme unifying fundamental interactions with gravity, such as superstrings, supergravity, or Grand-Unified Theories (GUTs). Moreover, higher-order curvature invariants emerge as loop corrections in high-curvature regimes, when matter fields are quantized on curved space-times (the result is that interactions between quantum fields and background geometry, or gravitational[5] self-interactions give rise to corrections in the Hilbert–Einstein Lagrangian). As shown in Ref. 18, these corrections are unavoidable in the effective quantum gravity actions, leading to a generalization of GR.[b] Of course, these models do not constitute a self-consistent quantum gravity theory, but are useful working schemes toward the construction of it.

---

[b]Higher-order curvature invariants like R^2, $R^{\mu\nu}R_{\mu\nu}$, $R^{\mu\nu\alpha\beta}R_{\mu\nu\alpha\beta}$, $R\,\Box R$, $R\,\Box^k R$, or non-minimal couplings between matter fields and geometry such as $\phi^2 R$ have to be added to the gravitational Lagrangian as soon as quantum corrections are introduced. For example, these terms occur in the low-energy limit of string Lagrangian or in Kaluza-Klein theories where extra spatial dimensions are taken into account.[19]

Besides these fundamental physics motivations, ETGs play a crucial role in cosmology because they exhibit inflationary behaviors able to overcome shortcomings of the standard cosmological model.[20,21]

An interesting aspect of higher-order and non-minimally coupled terms is that they can be transformed, via conformal transformations, to Einstein theory with one or more than one scalar field(s) minimally coupled to the curvature.[22-24] Specifically, in the Jordan frame, a term like R^2 in the Lagrangian gives fourth-order field equations, $R \Box R$ gives sixth-order equations,[25] $R \Box^2 R$ yields eighth-order equations,[26] and so on. Conformal transformations allow to get second-order derivative terms corresponding to a scalar field, therefore fourth-order gravity is conformally equivalent to Einstein gravity plus a scalar field; sixth-order gravity is conformally equivalent to GR plus two scalar fields; and so on.[27]

As mentioned before, ETGs play a fundamental role also in today's observed Universe. In fact, the increasing amount of observational data has given rise to a new cosmological model, the so-called *Concordance Model* or Λ-Cold Dark Matter (ΛCDM) model. The Hubble diagram of type Ia Supernovae (hereafter SNeIa) was the first evidence that the Universe is today undergoing an accelerated expansion phase. These data suggest that the Universe is dominated by an un-clustered fluid, with negative pressure, usually dubbed *dark energy*. Such a fluid drives the accelerated expansion. The ΛCDM model fails in explaining why the value of Λ is so tiny (120 orders of magnitude lower) if compared with the typical vacuum energy density predicted by particle physics, and why its present value is comparable to the matter density. This constitutes the so-called *coincidence problem*. Possible solutions are the models known as *quintessence*[28,29] (the origin of this quintessence scalar field is one of the big mysteries of modern cosmology), Chaplying gas,[30] tachyon fields,[31] condensate cosmology,[32] Cardassian model,[33] and Dvali–Gabadadze–Porrati (DGP) cosmology,[34] or non-vanishing torsion fields.[35]

In general, in the case of weak-field limit, which essentially coincides with Solar System scales, ETGs are expected to reproduce GR which is precisely tested at these scales.[1] Even this limit is a matter of debate because several theories do not reproduce exactly the Einstein theory in the Newtonian limit but, in some sense, generalize it giving rise to Yukawa-like corrections to the Newtonian potential which could be physically relevant already at Galactic scales.[36-41]

As a general remark, relativistic gravity theories give rise to corrections to the weak-field gravitational potentials which, at the post-Newtonian level and in the Parameterized Post-Newtonian (PPN) formalism, constitute an important test bed.[1] Furthermore, the *gravitational lensing astronomy*[42] provides additional tests over small, large, and very large scales which can provide measurements on the variation of Newton constant,[43] the potential of galaxies, clusters of galaxies, and other features of gravitating systems. In principle, such data are capable of confirming or ruling out alternatives to GR.

2.2. *Quantum Gravity motivations*

A challenge of modern physics is constructing a theory capable of describing the fundamental interactions of Nature under the same standard. This goal has led to formulate several unification schemes which attempt to describe gravity together with the other interactions. All these schemes describe fields under the conceptual apparatus of Quantum Mechanics. This is based on the assumption that the states of physical systems are described by vectors in a Hilbert space $\mathcal{H}$ and the physical fields are linear operators defined on domains of $\mathcal{H}$. Till now, any attempt to incorporate gravity into this scheme has failed or revealed to be unsatisfactory. The main problem is that gravitational field describes, at the same time, the gravitational degrees of freedom and the space-time background where these degrees of freedom are defined.

Owing to the difficulties of building up a self-consistent theory unifying interactions and particles, the two fundamental theories of modern physics, GR and Quantum Mechanics, have to be critically re-analyzed. On the one hand, we assume that matter fields (bosons and fermions) come out from superstructures (e.g. Higgs bosons or superstrings) that, undergoing certain phase transitions, generate the known particles. On the other hand, it is assumed that geometry interacts directly with quantum matter fields which back-react on it. This interaction necessarily modifies the standard gravitational theory, that is the Hilbert–Einstein Lagrangian. These facts lead to the ETGs as possible effective approaches toward Quantum Gravity.

From the point of view of cosmology, the modifications of GR provide inflationary scenarios of remarkable interest. In any case, a condition that such theories have to respect in order to be physically acceptable is that GR is recovered in the low-energy limit.

Although conceptual progresses have been made assuming generalized gravitational theories, at the same time mathematical difficulties have increased. The corrections into the Lagrangian enlarge the (intrinsic) non-linearity of the Einstein equations, making them more difficult to study because differential equations of higher order than second are often obtained and because it is extremely difficult to separate geometry from matter degrees of freedom. To overcome these difficulties and to simplify the field equations, one often looks for symmetries of dynamics and identifies conserved quantities which, often, allow to find out exact solutions.

The necessity of quantum gravity was recognized at the end of 1950s, when physicists tried to deal with all interactions at a fundamental level and describe them under the standard of quantum field theory. The first attempts to quantize gravity adopted both the canonical approach and the covariant approach, which had been already applied with success to electromagnetism. In the first approach, applied to electromagnetism, one takes into account electric and magnetic fields satisfying the Heisenberg Uncertainty Principle and the quantum states are gauge-invariant functionals, generated by the vector potential, defined on 3-surfaces labeled with constant time. In the second approach, one quantizes the two degrees of freedom of the Maxwell field without $(3+1)$ decomposition of the metric, while the quantum states are elements of a Fock space.[44] These procedures are fully equivalent. The former allows a well-defined measure, whereas the latter is more convenient for perturbative calculations such as the computation of the S-matrix in Quantum Electrodynamics (QED).

These methods have been adopted also in GR, but several difficulties arise due to the fact that GR cannot be formulated as a quantum field theory on a fixed Minkowski background. To be specific, in GR the geometry of background space-time cannot be given *a priori*: space-time is itself the dynamical variable. To introduce the notions of causality, time, and evolution, one has to solve equations of motion and build up the related space-time. For example, to know if a particular initial condition will give rise to a black hole, it is necessary to evolve it by solving the Einstein equations. Then, taking into account the causal structure of the solution, one has to study the asymptotic metric at future null infinity in order to assess whether it is related, in the causal past, with that initial condition. This problem cannot be handled at quantum level. Due to the Uncertainty Principle, in non-relativistic Quantum Mechanics, particles do not move along well-defined trajectories and one can only calculate the probability amplitude $\psi(t, x)$ that a measurement at a given time t detects a particle at the spatial point x. In the same way, in Quantum Gravity, the evolution of an initial state does not provide a given space-time (that is a metric).

In absence of a space-time, how is it possible to introduce basic concepts as causality, time, scattering matrix, or black holes?

Canonical and covariant approaches provide different answers to these issues. The first is based on the Hamiltonian formulation of GR and is adopting a canonical quantization procedure. The canonical commutation relations are those that lead to the Heisenberg Uncertainty Principle; in fact, the commutation of operators on a spatial three-manifold at constant time is assumed, and this three-manifold is fixed in order to preserve the notion of causality. In the limit of asymptotically flat space-time, motions related to the Hamiltonian have to be interpreted as time evolution (in other words, as soon as the background becomes the Minkowski space-time, the Hamiltonian operator assumes again its role as the generator of time translations). The canonical approach preserves the geometric structure of GR without introducing perturbative methods.

On the other hand, the covariant approach adopts QFT concepts. The basic idea is that the shortcomings mentioned above can be circumvented by splitting the metric $g_{\mu\nu}$ into a kinematic part $\eta_{\mu\nu}$ (usually flat) and a dynamical part $h_{\mu\nu}$. That is

$$g_{\mu\nu} = \eta_{\mu\nu} + h_{\mu\nu}. \tag{1}$$

The background geometry is given by the flat metric tensor and it is the same as the Special Relativity and standard QFT. It allows to define concepts of causality, time, and scattering. The procedure of quantization is applied to the dynamical field, considered as a little deviation of the metric from the flat background. Quanta are particles with spin two, i.e. *gravitons*, which propagate in Minkowski space-time and are defined by $h_{\mu\nu}$. Substituting $g_{\mu\nu}$ into the GR action, it follows that the gravitational Lagrangian contains a sum whose terms contain different orders of approximation, the interaction of gravitons and, eventually, terms describing matter–graviton interaction (if matter is present). These terms are analyzed by the standard perturbation approach of QFT.

These quantization approaches were both developed during the 1960s and 1970s. In the canonical approach, Arnowitt, Deser, and Misner developed the Hamiltonian formulation of GR using methods proposed by Dirac and Bergmann. In this scheme, the canonical variables are the 3-metric on the spatial three-manifolds obtained by foliating the four-dimensional manifold. It is worth noting that this foliation is arbitrary. Einstein's field equations give constraints between the three-metrics and their conjugate momenta and the evolution equation for these fields is the so-called

Wheeler–DeWitt (WDW) *equation*. In this picture, GR is the dynamical theory of the three-geometries, that is the *geometrodynamics*. The main problems arising from this formulation are that the quantum equations involve products of operators defined at the same space-time point and, furthermore, they give rise to distributions whose physical meaning is unclear. In any case, the main question is the absence of the Hilbert space of states and, as a consequence, the probabilistic interpretation of the quantities is not exactly defined.

The covariant quantization is much similar to the physics of particles and fields, because, in some sense, it has been possible to extend QED perturbation methods to gravitation. This allows the analysis of mutual interaction between gravitons and of the matter–graviton interactions. The Feynman rules for gravitons and the demonstration that the theory might be, in principle, unitary at any order of expansion was achieved by DeWitt.

Further progress was reached by the Yang–Mills theories, describing the strong, weak, and electromagnetic interactions of particles by means of symmetries. These theories are renormalizable because it is possible to give the particle masses through the mechanism of spontaneous symmetry breaking. According to this principle, it is natural to try to consider gravitation as a Yang–Mills theory in the covariant perturbation approach and search for its renormalization. However, gravity does not fit into this scheme; it turns out to be non-renormalizable if we consider the graviton–graviton interactions (two-loops diagrams) and graviton-matter interactions (one-loop diagrams).[c] In any case, the covariant method allows to construct a gravity theory which is renormalizable at one-loop level in the perturbation series.[5]

Due to the non-renormalizability of gravity at higher orders, the validity of the approach is restricted to the low-energy domain, that is, to infrared scales, while it fails at ultraviolet scales. This implies that the theory of gravity is unknown near or at Planck scales. On the other hand, sufficiently far from the Planck epoch, GR and its first loop approximation describe gravitation quite well. In this context, it makes sense to add higher-order and non-minimally coupled terms to the Hilbert–Einstein action. Furthermore, if the free parameters are chosen appropriately, the theory has a

---

[c]Higher order terms in the perturbative series imply an infinite number of free parameters. At the one-loop level, it is sufficient to renormalize only the effective constants G_{eff} and Λ_{eff} which, at low energy, reduce to the Newton constant G_N and the cosmological constant Λ.

better ultraviolet behavior and it is asymptotically free. Nevertheless, the Hamiltonian of these theories is not bounded from below and they are unstable. Specifically, unitarity is violated and probability is not conserved.

Another approach to Quantum Gravity comes from the electroweak interaction. Here, gravity is treated neglecting the other fundamental interactions. The unification of electromagnetism and weak interaction suggests that it could be possible to obtain a consistent theory if gravity is coupled to some kind of matter. This is the basic idea of Supergravity. In this kind of theories, divergences due to bosons (in this case the 2-spin gravitons) are cancelled exactly by those due to the fermions. In this picture, it is possible to achieve a renormalized theory of gravity. Unfortunately, the approach works only up to two-loop level and for matter-gravity couplings. The corresponding Hamiltonian is positive-definite and the theory is unitary. However, including higher-order loops, the infinities appear and renormalizabilty is lost.

Perturbation methods are also adopted in String Theory. In this case, the approach is different from the previous one since particles are replaced by extended objects which are the fundamental strings. The physical particles, including the spin two gravitons, correspond to excitations of the strings. The only free parameter of the theory is the string tension and the interaction couplings are uniquely determined. As a consequence, string theory contains all fundamental physics and it is considered a possible *Theory of Everything*. String theory is unitary and the perturbation series converges implying finite terms. This feature follows from the characteristic that strings are intrinsically extended objects, so that ultraviolet divergencies, appearing at small scales or at large transferred impulses, are naturally cured. This means that the natural cutoff is given by the string length, which is of Planck size l_P. At larger scales than l_P, the effective string action can be written as non-massive vibrational modes, that is, in terms of scalar and tensor fields. This constitutes the *tree-level effective action*. This approach leads to an effective theory of gravity non-minimally coupled with scalar fields, which are the so-called *dilaton fields*.

In conclusion, we can summarize the previous considerations: (1) a unitary and renormalizable theory of gravity does not exist yet[d]. (2) In

---

[d]It is worth to mention that it has been shown that an infinite derivative theory of covariant gravity, which is motivated from string theory, see Refs. 45, 46, can be made ghost free and also singularity free[47, 48] (see Refs. 49–53 for some applications).

the quantization program of gravity, two approaches are used: the *covariant approach* and the *perturbation approach*. They do not lead to a self-consistent Quantum Gravity. (3) In the low-energy regime, with respect to the Planck energy, GR can be improved by introducing, into the Hilbert–Einstein action, higher-order terms of curvature invariants and non-minimal couplings between matter and gravity. The approach leads, at least at one-loop level, to a consistent and renormalizable theory.

In all these considerations, the fact that EP has to hold or can be violated plays a fundamental role. possible violations at quantum level could allow to deal with gravity under the same standard of other interactions.

2.3. *Quantum field theory in curved space-time*

Any attempt to formulate QFT on curved space-time necessarily leads to modifying the Hilbert–Einstein action. This means adding terms containing non-linear invariants of the curvature tensor or non-minimal couplings between matter and curvature. The approach could led to the violation of EP.

At high energies, a description of matter as a hydrodynamical perfect fluid is inadequate: an accurate description asks for QFT formulated on a curved space-time. Since, at scales comparable to the Compton wavelength of particles, matter has to be quantized, one can adopt a semiclassical description of gravity where the Einstein field equations assume the form

$$G_{\mu\nu} \equiv R_{\mu\nu} - \frac{1}{2} g_{\mu\nu} R = \langle T_{\mu\nu} \rangle, \qquad (2)$$

where the Einstein tensor $G_{\mu\nu}$ is on the left-hand side while the right-hand contains the expectation value of quantum stress–energy tensor which is the source of the gravitational field. Specifically, if $|\psi>$ is a quantum state, then $\langle T_{\mu\nu} \rangle \equiv \langle \psi | \hat{T}_{\mu\nu} | \psi \rangle$, where $\hat{T}_{\mu\nu}$ is the quantum operator associated with the classical energy–momentum tensor of the matter field with a regularized expectation value.

If the background is curved, quantum fluctuations of matter fields give, even in absence of classical matter and radiation, non-vanishing contributions to $< T_{\mu\nu} >$ as in QED.[5] If matter fields are free, massless, and conformally invariant, these corrections are

$$\langle T_{\mu\nu} \rangle = k_1 \,^{(1)} H_{\mu\nu} + k_3 \,^{(3)} H_{\mu\nu}. \qquad (3)$$

Here, $k_{1,3}$ are numerical coefficients and

$$^{(1)}H_{\mu\nu} = 2R_{;\mu\nu} - 2g_{\mu\nu}\Box R + 2R^{\sigma\tau}R_{\sigma\tau}g_{\mu\nu} - \frac{1}{2}g_{\mu\nu}R^2, \tag{4}$$

$$^{(3)}H_{\mu\nu} = R^\sigma{}_\mu R_{\nu\sigma} - \frac{2}{3}RR_{\mu\nu} - \frac{1}{2}g_{\mu\nu}R^{\sigma\tau}R_{\sigma\tau} + \frac{1}{4}g_{\mu\nu}R^2. \tag{5}$$

$^{(1)}H_{\mu\nu}$ is a tensor derived by varying the local action,

$$^{(1)}H_{\mu\nu} = \frac{2}{\sqrt{-g}}\frac{\delta}{\delta g^{\mu\nu}}\left(\sqrt{-g}\,R^2\right). \tag{6}$$

It is divergence free, that is $^{(1)}H^\nu_{\mu;\nu} = 0$. Infinities coming from $\langle T_{\mu\nu}\rangle$ are removed by introducing an infinite number of counterterms in the Lagrangian density of gravitation. The procedure yields a renormalizable theory. For example, one of these terms is $CR^2\sqrt{-g}$, where C indicates a parameter that diverges logarithmically. Equation (2) cannot be generated by a finite action because the gravitational field would be completely renormalizable, that is, it would eliminate a finite number of divergences to make gravitation similar to QED. On the contrary, one can only construct a truncated quantum theory of gravity. The parameter used for the expansion in loop is the Planck constant $\hbar$. It follows that the truncated theory at the one-loop level contains all terms of order $\hbar$, that is the first quantum correction. Some points have to be stressed now: (1) Matter fields are *free* and, if the Equivalence Principle is valid at quantum level, all forms of matter couple in the same way to gravity. (2) The intrinsic non-linearity of gravity naturally arises, and then a number of loops are necessary to take into account self-interactions between matter and gravitation. In view of removing divergences at one-loop order, one has to renormalize the gravitational coupling G_{eff} and the cosmological constant Λ_{eff}. One-loop contributions of $\langle T_{\mu\nu}\rangle$ are the quantities introduced above, that is $^{(1)}H_{\mu\nu}$ and $^{(3)}H_{\mu\nu}$. Furthermore, one has to consider

$$^{(2)}H_{\mu\nu} = 2R^\sigma{}_{\mu;\nu\sigma} - \Box R_{\mu\nu} - \frac{1}{2}g_{\mu\nu}\Box R + R^\sigma{}_\mu R_{\sigma\nu} - \frac{1}{2}R^{\sigma\tau}R_{\sigma\tau}g_{\mu\nu}. \tag{7}$$

In a conformally flat space-time, one has $^{(2)}H_{\mu\nu} = \frac{1}{3}{}^{(1)}H_{\mu\nu}$,[5] so that only the first and third terms of $H_{\mu\nu}$ are present in (3).

The tensor $^{(3)}H_{\mu\nu}$ is conserved only in conformally flat space-times and it cannot be obtained by varying a local action. The trace of the energy–momentum tensor is null for conformally invariant classical fields, while one finds that the expectation value of the tensor (3) has non-vanishing trace. This result gives rise to the so-called *trace anomaly*.[5]

By summing all the geometric terms in $\langle T^\rho_\rho \rangle_{ren}$, deduced by the Riemann tensor and of the same order, one derives the right-hand side of (3). In the case in which the background metric is conformally flat, it can be expressed in terms of Eqs. (4) and (5). We conclude that the trace anomaly, related to the geometric terms, emerges because of the one-loop terms due to QFT on curved space-time.[e]

Masses of the matter fields and their mutual interactions can be neglected in the high-curvature limit because $R \gg m^2$. On the other hand, matter–graviton interactions generate non-minimal couplings in the effective Lagrangian. The one-loop contributions of such terms are comparable to those given by the trace anomaly and generate, by conformal transformations, the same effects on gravity.

The simplest effective Lagrangian taking into account these corrections is

$$\mathcal{L}_{NMC} = -\frac{1}{2}\nabla^\alpha\varphi\nabla_\alpha\varphi - V(\varphi) - \frac{\xi}{2}R\phi^2, \qquad (8)$$

where ξ is a dimensionless constant. The stress-energy tensor of the scalar field results modified accordingly, but a conformal transformation can be found such that the modifications related to curvature terms can be cast in the form of a matter-curvature interaction. The same argument holds for the trace anomaly. Some Grand Unification Theories lead to polynomial couplings of the form $1 + \xi\phi^2 + \zeta\phi^4$ that generalize the one in (8). An exponential coupling $e^{-\alpha\varphi}R$ between a scalar field (dilaton) φ and the Ricci scalar appears in the effective Lagrangian of strings.

Field equations derived by varying the action $\mathcal{L}_{NMC}$ are

$$\left(1 - \kappa\xi\phi^2\right)G_{\mu\nu} = \kappa\left\{\nabla_\mu\phi\nabla_\nu\phi - \frac{1}{2}g_{\mu\nu}\nabla^\alpha\phi\nabla_\alpha\phi - Vg_{\mu\nu}\right.$$

$$\left. +\xi\left[g_{\mu\nu}\Box\left(\phi^2\right) - \nabla_\mu\nabla_\nu\left(\phi^2\right)\right]\right\}, \qquad (9)$$

$$\Box\phi - \frac{dV}{d\phi} - \xi R\phi = 0. \qquad (10)$$

---

[e]Equations (4) and (5) can include terms containing derivatives of the metric of order higher than fourth (fourth order being the R^2 term) if all possible Feynman diagrams are included. For example, corrections such as $R\Box R$ or $R^2\Box R$ can be present in $^{(3)}H_{\mu\nu}$ implying equations of motion that contain sixth-order derivatives of the metric. Also, these terms can be treated by making use of conformal transformations.[25]

The non-minimal coupling of the scalar field is similar to that derived for the 4-vector potential of curved space in Maxwell theory.

Several motivations can be provided for the non-minimal coupling in the Lagrangian $\mathcal{L}_{NMC}$. A nonzero ξ is generated by first loop corrections even if it does not appear in the classical action.[5,54–57] Renormalization of a classical theory with $\xi = 0$ shifts this coupling constant to a value which is small.[58,59] It can, however, affect drastically an inflationary cosmological scenario and determine its success or failure.[60–63] A non-minimal coupling is expected at high curvatures.[56,57] Furthermore, non-minimal coupling solves potential problems of primordial nucleosynthesis[64] and, besides, the absence of pathologies in the propagation of φ-waves requires conformal coupling for all non-gravitational fields.[65–69,f]

The conformal value $\xi = 1/6$ is fixed at infrared scales of renormalization group.[71–76] Non-minimally coupled scalar fields have been used in inflationary scenarios.[77–86] The approach adopted was considering ξ as a free parameter to fix problems of specific inflationary scenarios.[63,87] Cosmological reheating with strong coupling $\xi \gg 1$ has also been studied[81,88,89] and considered in relation with wormholes,[90–92] black holes,[93,94] and boson stars.[95–97] The coupling ξ is not, in general, a free parameter but depends on the particular scalar field φ considered.[56,57,62,63,87,98–100] As we will see in what follows, the coupling with a scalar field could be a mechanism leading to possible violations of EP when theories at finite temperature are considered.

2.4. *Some relevant aspects of the Equivalence Principle*

The above motivations point out the necessity of a critical analysis of EP and its possible violations.

As we mentioned, the starting point for any formulation of the EP is the equivalence between inertial and gravitational mass $m_I = m_G$ (Galileo's experiment), which implies that all bodies fall with the same acceleration, independently of their mass and internal structure, in a given gravitational field (universality of free fall or WEP). A more precise statement of WEP is[1]

---

[f]Note, however, that the distinction between gravitational and non-gravitational fields becomes representation-dependent in ETGs, together with the various formulations of the EP.[70]

If an uncharged body is placed at an initial event in space-time and given an initial velocity there, then its subsequent trajectory will be independent of its internal structure and composition

This formulation of WEP was enlarged by Einstein adding a new fundamental part. According to him, in a local inertial frame (the free-falling elevator) not only do the laws of mechanics behave in it as if gravity were absent, but *all* physical laws (except those of gravitational physics) have the same behavior. The current terminology refers to this principle as the *Einstein Equivalence Principle* (EEP). A more precise statement is[1]

"The outcome of any local non-gravitational test experiment is independent of the velocity of the (free-falling) apparatus and the outcome of any local non-gravitational test experiment[g] *is independent of where and when in the Universe it is performed"*.

From the EEP, it follows that the gravitational interaction must be described in terms of a curved space-time, that is the postulates of the so-called metric theories of gravity have to be satisfied[1]:

(1) space-time is endowed with a metric $g_{\mu\nu}$;
(2) the world lines of test bodies are geodesics of that metric;
(3) in local freely falling frames (called *local Lorentz frames*), the non-gravitational laws of physics are those of Special Relativity.

These definitions characterize the most fundamental feature of GR, hence the EP, as well as the physical properties that allow to discriminate between GR and other metric theories of gravity. In the ETGs, some additional features arise because these definitions depend on the conformal representation of the adopted theory. More precisely, in scalar-tensor gravity, massive test particles in the Jordan frame follow geodesics, satisfying the WEP, but the same particles deviate from geodesic motion in the Einstein frame. This difference shows that the EP is formulated in a representation-dependent way.[70] This serious shortcoming has not yet been addressed properly.

Some specific features, related to the Equivalence Principle, can be considered:

---

[g]A "local non-gravitational experiment" is defined as an experiment performed in a small size freely falling laboratory, in order to avoid the inhomogeneities of the external gravitational field, and in which any gravitational self-interaction can be ignored. For example, the measurement of the fine structure constant is a local non-gravitational experiment, while the Cavendish experiment is not.

- Let us assume that WEP is violated. Let us assume, for example, that the inertial masses (m_{Ii}) in a system differ from the passive ones,

$$m_{Pi} = m_{Ii}\left(1 + \Sigma_A \eta^A \frac{E^A}{m_{Ii}c^2}\right),$$ (11)

where E^A is the internal energy of the body connected to the A-interaction and η^A is a dimensionless parameter quantifying the violation of the WEP. It is then convenient to introduce a new dimensionless parameter (the *Eötvös ratio*) considering, for example, two bodies moving with accelerations

$$a_i = \left(1 + \Sigma_A \eta^A \frac{E^A}{m_{Ii}c^2}\right) g \quad (i = 1, 2);$$ (12)

where g is now the gravitational acceleration. Then we define the Eötvos ratio as

$$\eta = 2\frac{|a_1 - a_2|}{|a_1 + a_2|} = \Sigma_A \eta^A \left(\frac{E_1^A}{m_{I1}c^2} - \frac{E_2^A}{m_{I2}c^2}\right).$$ (13)

The measured value of η provides information on the WEP-violation parameters η^A. Experimentally, the equivalence between inertial and gravitational masses is strongly confirmed.[1]
- The minimal coupling prescription is also important since it is connected with the mathematical formulation of the EEP (actually, to implement the EEP one needs to put in special-relativistic form the laws under consideration and then proceed to find the general-relativistic formulation, switching on gravity. In other words, we have to apply minimal coupling prescriptions with the caveat already discussed).
- The last point is strictly related to the scalar-tensor theories of gravity: do these theories satisfy the EEP?

    To address this question, one has to generalize the above two principles and introduce new concepts. Following Will,[1] one introduces the notion of "purely dynamical metric theory", i.e. a theory in which *the behavior of each field is influenced to some extent by a coupling to, at least, one of the other fields in the theory.*[1] In this respect, GR is a purely dynamical theory, as well as the Brans–Dicke theory since the equations for the metric involve the scalar field, and vice versa.

In these theories, the calculations of the metric are done in two stages: (1) the assignment of boundary conditions "far" from the local system; (2) inferring the solutions of equations for the fields generated by the local system. Owing to the coupling of metric with fields (for given boundary conditions), the latter will influence the metric. This implies that local gravitational experiments can depend on where the lab is located in the Universe, as well as on its velocity relative to the external world. One of the consequences of such a new physical scenario is that, in a Brans–Dicke theory, and more generally in Scalar–Tensor Theories, the gravitational coupling "constant" turns out to depend on the asymptotic value of the scalar field.

All these considerations are strictly related to the *Strong Equivalence Principle* (SEP)[1]:

(i) "WEP is valid for self-gravitating bodies as well as for test bodies;
(ii) the outcome of any local test experiment is independent of the velocity of the (freely falling) apparatus;
(iii) the outcome of any local test experiment is independent on where and when in the Universe it is performed".[1]

The SEP differs from the EEP because it includes the self-gravitating interactions of bodies (such as planets or stars), and because of experiments involving gravitational forces (e.g., the Cavendish experiment). SEP reduces to the EEP when gravitational forces are ignored. Many authors have conjectured that the only theory compatible with the SEP is GR (that is $SEP \longrightarrow GR -$ only).

2.5. *The Schiff conjecture*

The Schiff conjecture represents one of the most important topics related to the foundations of the gravitational physics. Its original formulation asserts that *every theory of gravity that satisfies WEP and is relativistic, necessarily satisfies the EEP, and is consequently a metric theory of gravity.* Hence, $WEP \Rightarrow EEP$. Later, Will proposed a slight modification of Schiff conjecture: *every theory of gravity that satisfies WEP and the principle of universality of gravitational red-shift (UGR) necessarily satisfies EEP.* Hence, in such a case, $WEP + UGR \rightarrow EEP$.

Let us discuss in some details these topics. Note that the correctness of Schiff's conjecture implies that the Eötvös and the gravitational red-shift experiments would provide a direct empirical confirmation of the EEP, with the consequence that gravity can be interpreted as a geometrical (curved space-time) phenomenon. The relevance of such a fundamental aspect of the gravitational physics led to different mathematical approaches to prove the Schiff conjecture. These frameworks encompass all metric theories, as well as non-metric theories of gravity. Lightman and Lee[101, 102] proved Schiff's conjecture in the framework of the so-called $TH\epsilon\mu$ formalism. They consider the motion of charged particles (electromagnetic coupling) in a static spherically symmetric gravitational field $U = GM/r$

$$S_{TH\epsilon\mu} = -\sum_a m_a \int dt \sqrt{T - Hv_a^2} + \sum_a e_a \int dt v_a^\mu A_\mu(x_a^\mu)$$
$$+ \frac{1}{2} \int d^4x \left(\epsilon \mathbf{E}^2 + \frac{\mathbf{B}^2}{\mu} \right), \tag{14}$$

where $m_a, e_a, v_a^\mu \equiv \dfrac{dx_a^\mu}{dt}$ represent the mass, the charge, and the velocity of the particle a, respectively. The parameters $TH\epsilon\mu$ do depend on the gravitational field U, that is they essentially account for the response of the electromagnetic fields to the external potential, and may vary from theory to theory. A metric theory must satisfy the relation

$$\epsilon = \mu = \sqrt{\frac{H}{T}}, \quad \text{for all } U.$$

In the case of non-metric theories, the parameters $TH\epsilon\mu$ may depend on the species of particles or on the field coupling to gravity. The metric is given by $ds^2 = T(r)dt^2 - H(r)(dr^2 + r^2 d\Omega)$. Lightmann and Lee showed in Ref. 101 that the rate of fall of a test body made up of interacting charged particles does not depend of the structure of the body (WEP) *if and only if* $\epsilon = \mu = \sqrt{\dfrac{H}{T}}$. This implies $WEP \Rightarrow EEP$, satisfying hence the Schiff conjecture.

Will generalized the Dirac equation in $TH\epsilon\mu$ formalism, and computed the gravitational red-shift experienced by different atomic clocks showing that the red-shift is independent of the nature of clocks (Universality of Gravitational Red-shift (UGR)) *if and only if* $\epsilon = \mu = \sqrt{\dfrac{H}{T}}$.[103] Therefore $UGR \Rightarrow EEP$, verifying in such a way another aspect of the Schiff conjecture.[104]

W. T. Ni was able to provide a counterexample to Schiff's conjecture by considering the coupling between a pseudo-scalar field ϕ with the electromagnetism field

$$\mathcal{L}_{\phi F} \sim \phi \varepsilon^{\alpha\beta\gamma\delta} F_{\alpha\beta} F_{\gamma\delta}, \tag{15}$$

where $\varepsilon^{\alpha\beta\gamma\delta}$ is the completely anti-symmetric Levi-Civita symbol.[105] In Refs. 106–108, the Schiff conjecture is analyzed in the framework of gravitational non-minimally coupled theories. More specifically, the total Lagrangian density considered is given by

$$\mathcal{L}_{NMC} = \frac{R}{16\pi G} + \mathcal{L}_M + \mathcal{L}_I(\psi^A, g_{\mu\nu}), \tag{16}$$

where $\mathcal{L}_I(\psi^A, g_{\mu\nu})$ is the Lagrangian density of some field ψ^A non-minimally coupled to gravity,[107, 108] while $\mathcal{L}_I = \chi^{\alpha\beta\gamma\delta} R_{\alpha\beta\gamma\delta}$ in,[106] where $\chi^{\alpha\beta\gamma\delta}$ depends on matter, for example

$$\chi^{\alpha\beta\gamma\delta} = \bar{\psi}\sigma^{\alpha\beta}\psi\bar{\psi}\sigma^{\gamma\delta}\psi, \psi^{\alpha\mu}\psi^{\beta\nu} - \psi^{\beta\mu}\psi^{\alpha\nu}, \tag{17}$$

with ψ being a spin-half field and $\psi^{\alpha\beta}$, a (non-gravitational) spin-2 field. Both results show that these gravitational theories are, in general, incompatible with Schiff's conjecture.

These counterexamples indicate that a rigorous proof of such a conjecture is impossible. However, some powerful arguments of plausibility can be formulated. One of them is based upon the assumption of energy conservation.[109] Following,[110] let us consider a system in a quantum state $|A\rangle$ that decays in a state $|B\rangle$, with the emission of a photon with frequency ν. The quantum system falls from a height H in an external gravitational field $gH = \Delta U$, so that the system in state $|B\rangle$ falls with acceleration g_B and the photon frequency is shifted to ν'. Assuming a violation of the WEP, the acceleration g_A and g_B of the system $|A\rangle$ and $|B\rangle$ are

$$g_A = g\left(1 + \frac{\alpha E_A}{m_A}\right), \quad g_A = g\left(1 + \frac{\alpha E_A}{m_A}\right), \quad E_B - E_A = h\nu \tag{18}$$

that is they depend on that portion of the internal energy of the states. Here ν is the frequency of the quantum emitted by the system $|A\rangle$. The conservation of energy implies that there must be a corresponding violation of local position invariance in the frequency shift given by $\frac{\nu'-\nu}{\nu} = (1+\alpha)\Delta U$, where ν' is the frequency of the quantum at the bottom of the trajectory.

The Eötvös parameter is (for $m_A \sim m_B \sim m$)

$$\eta = \frac{|g_B - g_A|}{|g_B + g_A|} \simeq \frac{\alpha(E_A - E_B)}{m}. \tag{19}$$

The Schiff conjecture is still nowadays a subject of a strong scientific debate and deep scrutiny.

2.6. Mach's principle and the variation of Newton constant

The EP and the Mach principle can also be deeply connected. We can discuss this point with some very straightforward considerations.

Following Bondi[14] there are, at least in principle, two entirely different ways of measuring the rotational velocity of Earth. The first is a purely terrestrial experiment (e.g., a Foucault pendulum), while the second is an astronomical observation consisting of measuring the terrestrial rotation with respect to the fixed stars. In the first type of experiment, the motion of the Earth is referred to as an idealized inertial frame in which Newton's laws are verified. However, a unique general relativistic approach to define rotations has been introduced by Pirani considering the boucing photons[111,112] (see also Ref. 113).

In the second kind of experiment, the frame of reference is connected to a matter distribution surrounding the Earth and the motion of the latter is referred to this matter distribution. In this way, we face the problem of Mach's principle, which essentially states that the local inertial frame is determined by some average motion of distant astronomical objects.[14,17,h] Trying to incorporate Mach's principle into metric gravity, Brans and Dicke constructed a theory alternative to GR.[15] Taking into account the influence that the total matter has at each point (constructing the "inertia"), these two authors introduced, together with the standard metric tensor, a new scalar field of gravitational origin as the effective gravitational coupling. This is why the theory is referred to as a "scalar-tensor" theory; actually, theories in this spirit had already been proposed years earlier by Jordan, Fierz, and Thiery (see Ref. 115). An important ingredient of this approach is that the gravitational "constant" is actually a function of the total mass

---

[h]An interesting discussion on this topic, also connected with different theories of space, both in philosophy and in physics, is found in Dicke's contribution "The Many Faces of Mach" in Gravitation and Relativity.[114] This discussion presents also the problematic positions that Einstein had on Mach's principle.

distribution, that is of the scalar field, and it is actually variable. In this picture, gravity is described by the Lagrangian density

$$\mathcal{L}_{BD} = \sqrt{-g}\left[\varphi R - \frac{\omega}{\varphi}\nabla^\mu\varphi\nabla_\mu\varphi + \mathcal{L}^{(m)}\right], \tag{20}$$

where ω is the dimensionless Brans–Dicke parameter and $\mathcal{L}^{(m)}$ is the matter Lagrangian including all the non-gravitational fields. As stressed by Dicke,[116] the Lagrangian (20) has a property similar to one already discussed in the context of higher-order gravity. Under the conformal transformation $g_{\mu\nu} \to \tilde{g}_{\mu\nu} = \Omega^2 g_{\mu\nu}$ with $\Omega = \sqrt{G_0\varphi}$, the Lagrangian (20) is mapped into

$$\mathcal{L} = \sqrt{-\tilde{g}}\left(\tilde{R} + G_0\tilde{\mathcal{L}}^{(m)} + G_0\tilde{\mathcal{L}}^{(\Omega)}\right), \tag{21}$$

where

$$\tilde{\mathcal{L}}^{(\Omega)} = -\frac{(2\omega+3)}{4\pi G_0\Omega}(\nabla^\alpha\sqrt{\Omega})(\nabla_\alpha\sqrt{\Omega}), \tag{22}$$

and $\tilde{\mathcal{L}}^{(m)}$ is the conformally transformed Lagrangian density of matter. In this way, the total matter Lagrangian $\tilde{\mathcal{L}}_{tot} = \tilde{\mathcal{L}}^{(m)} + \tilde{\mathcal{L}}^{(\Omega)}$ has been introduced. The field equations are now written in the form of Einstein-like equations as

$$\tilde{R}_{\mu\nu} - \frac{1}{2}\tilde{g}_{\mu\nu}\tilde{R} = G_0\tilde{\tau}_{\mu\nu}, \tag{23}$$

where the stress–energy tensor is now the sum of two contributions,

$$\tilde{\tau}_{\mu\nu} = T^{(m)}_{\mu\nu} + \Lambda_{\mu\nu}(\Omega). \tag{24}$$

Dicke noted that this new (tilded, or Einstein frame) form of the scalar-tensor theory has certain advantages over the theory expressed in the previous (non-tilded, or Jordan frame) form; the Einstein frame representation, being similar to the Einstein standard description, is familiar and easier to handle in some respects. But, in this new form, the Brans–Dicke theory also exhibits unpleasant features. If we consider the motion of a spinless, electrically neutral, massive particle, we find that, in the conformally rescaled world, its trajectory is no longer a geodesic. Only null rays remain unchanged by the conformal rescaling. This is a manifestation of the fact that the rest mass is not constant in the conformally transformed world and the equation of motion of massive particles is modified by the addition of an extra force proportional to $\nabla^\mu\Omega$.[116] Photon trajectories, on the other hand, are not modified because the vanishing of the photon mass implies the absence of a preferred physical scale and photons stay massless under the conformal rescaling, therefore their trajectories are unaffected.

This new approach to gravitation has increased the relevance of theories with varying gravitational coupling. They are of particular interest in cosmology since they have the potentiality to circumvent many shortcomings of the Standard Cosmological Model. We list here the Lagrangians of this type which are most relevant for this review.

- The low-energy limit of the bosonic string theory[117–119] produces the Lagrangian

$$\mathcal{L} = \sqrt{-g}\, e^{-2\varphi} \left(R + 4g^{\mu\nu} \nabla_\mu \varphi \nabla_\nu \varphi - \Lambda \right). \tag{25}$$

- The general scalar-tensor Lagrangian is

$$\mathcal{L}_{ST} = \sqrt{-g} \left[f(\varphi) R - \frac{\omega(\varphi)}{2}\, g^{\alpha\beta} \nabla_\alpha \varphi \nabla_\beta \varphi - V(\varphi) \right], \tag{26}$$

where $f(\varphi)$ and $\omega(\varphi)$ are arbitrary coupling functions and $V(\varphi)$ is a scalar field potential. The original Brans–Dicke Lagrangian is contained as the special case $f(\varphi) = \varphi, \omega(\varphi) = \omega_0/\varphi$ (with ω_0 a constant), and $V(\varphi) \equiv 0$.
- A special case of the previous general theory is that of a scalar field non-minimally coupled to the Ricci curvature, which has received so much attention in the literature to deserve a separate mention,

$$\mathcal{L}_{NMC} = \sqrt{-g} \left[\left(\frac{1}{16\pi G} - \frac{\xi}{2} \right) R - \frac{1}{2} g_{\mu\nu} \nabla_\mu \varphi \nabla_\nu \varphi - V(\varphi) \right], \tag{27}$$

where ξ is a dimensionless non-minimal coupling constant. This explicit non-minimal coupling was originally introduced in the context of classical radiation problems[120] and, later, conformal coupling with $\xi = 1/6$ was discovered to be necessary for the renormalizability of the $\lambda\varphi^4$ theory on a curved space-time.[5, 121] As said, the theory is conformally invariant for $\xi = 1/6$ and either $V \equiv 0$ or $V = \lambda\varphi^{4}$[5, 121–123] are allowed potentials.

All these theories exhibit a non-constant gravitational coupling. The Newton constant G_N is replaced by the effective gravitational coupling

$$G_{eff} = \frac{1}{f(\varphi)} \tag{28}$$

in Eq. (26) which, in general, is different from G_N. In string theory, such functions are specified in (25). In particular, in spatially homogeneous and isotropic cosmology, the coupling G_{eff} can only be a function of the epoch, i.e. of the cosmological time.

We stress that all these scalar–tensor theories of gravity do not satisfy the SEP because of the above-mentioned feature: the variation of G_{eff} implies that local gravitational physics depends on the scalar field via φ. This fact motivates the introduction of a stronger version of EP, the SEP. General theories with such a peculiar aspect are called *non-minimally coupled theories*. This generalizes older terminology in which the expression "non-minimally coupled scalar" referred specifically to the field described by the Lagrangian $\mathcal{L}_{NMC}$ of (27), which is a special case of (26).

Let us consider, as in (26), a general scalar–tensor theory in presence of "standard" matter with total Lagrangian density $\varphi R + \mathcal{L}^{(\varphi)} + \mathcal{L}^{(m)}$, where $\mathcal{L}^{(m)}$ describes ordinary matter. The dynamical equations for this matter are contained in the covariant conservation equation $\nabla^\nu T_{\mu\nu}^{(m)} = 0$ for the matter stress-energy tensor $T_{\mu\nu}^{(m)}$, which is derived from the variation of the total Lagrangian with respect to $g^{\mu\nu}$. In other words: concerning standard matter, everything goes as in GR (i.e. $\eta_{\mu\nu} \to g_{\mu\nu}$, $\partial_\mu \to \nabla_\mu$) following the minimal coupling prescription. What is new in these theories is the way in which the scalar and the metric degrees of freedom appear: now there is a direct coupling between the scalar degree of freedom and a function of the tensor degree of freedom (the metric) and its derivatives (specifically, with the Ricci scalar of the metric $R\left(g, \partial g, \partial^2 g\right)$). Then, confining our analysis to the cosmological arena, we face two alternatives. The first is

$$\lim_{t \to \infty} G_{eff}\left(\varphi(t)\right) = G_N, \tag{29}$$

this is the case in which standard GR cosmology is recovered at the present time in the history of the Universe. The second possibility occurs if the gravitational coupling is not constant today, i.e. G_{eff} is still varying with the epoch and $\dot{G}_{eff}/G_{eff}|_{now}$ (in brief $\dot{G}/G$) is non-vanishing.

In many theories of gravity, then, it is perfectly conceivable that G_{eff} varies with time: in some solutions G_{eff} does not even converge to the value observed today. What do we know, from the observational point of view, about this variability? There are three main approaches to analyze the variability of G_{eff}: the first is the *lunar laser ranging* (LLR) monitoring the Earth–Moon distance; the second is the information from solar astronomy; the third consists of data from binary pulsars.

The LLR consists of measuring the round trip travel time and thus the distances between transmitter and reflector, and monitoring them over an extended period of time. The change of round trip time contains information about the Earth–Moon system. This round trip travel time has been measured for more than 25 years in connection with the Apollo 11, 14, 15, and the Lunakhod 2 lunar missions. Combining these data with those coming from the evolution of the Sun (the luminosity of main sequence stars is quite sensitive to the value of G) and the Earth-Mars radar ranging, the current bounds on $\dot{G}/G$ allow at most 0.4×10^{-11} to 1.0×10^{-11} per year.[124] Another source of information on G-variability is given by binary pulsar systems. In order to extract data from this type of systems (the prototype is the famous binary pulsar PSR 1913+16 of Hulse and Taylor[125]), it has been necessary to extend the post-Newtonian approximation, which can be applied only to a weakly (gravitationally) interacting n-body system, to strongly (gravitationally) interacting systems. The order of magnitude of $\dot{G}/G$ allowed by these strongly interacting systems is 2×10^{-11} yr^{-1}.[124]

A general remark is necessary at this point. According to the Mach Principle, gravity can be considered as an average interaction given by the distribution of celestial bodies. This means that the same gravitational coupling can be related to the space-time scale, then supposing a variation of G_N is an issue to make the theory more Machian. From an experimental point of view, this fact reflects on the uncertainties of the measurements of G_N and it could constitute a test for any alternative theory of gravity with respect to GR.

Finally it is worth noting that there also exist Higgs-scalar-tensor theories (see for example Refs. 126–128) where inertia and gravity are strongly related. Such theories have been introduced to solve the issues raised in the Brans–Dicke theory where the observational results, coming from the Mercury perihelion shift, are not matched. In view of this shortcoming, Dicke postulated the existence of a mass-quadrupole momentum giving rise to an oblateness correction of the Sun shape. Since this feature was not detected, Higgs–scalar–tensor theories were deemed necessary.

2.7. The Equivalence Principle in Poincaré Gauge gravity and torsion

As we have seen the EP sates, the effect of gravity on matter is locally equivalent to the effect of a non-inertial reference frame in Special Relativity. The dynamical content of the EP can be understood by considering an inertial frame in[i] M_4, in which a matter field ϕ is described by the Lagrangian $L_M(\phi; \partial_i \phi)$. Passing to a non-inertial frame, L_M transforms into $\sqrt{-g} L_M(\phi; \nabla_i \phi)$, with $\nabla_i = e_i^\mu (\partial_\mu + \omega_\mu)$ the covariant derivative. The gravitational field (equivalent to the non-inertial reference frame) appears in the quantities $\sqrt{-g}$ and ∇_i, and can be eliminated on the whole space-time by reducing to the global inertial frame, while real gravitational fields can be eliminated only locally. For introducing a real gravitational field, hence, Einstein replaced M_4 with a Riemann space V_4. However, also a Riemann–Cartan space U_4 can be chosen.[129]

Another formulation of the EP asserts that the effect of gravity on matter can be locally eliminated by a suitable choice of reference frame, and matter behaves according to the laws of Special Relativity,[129] i.e. *at any point P in space-time an orthonormal reference frame $\mathbf{e}_i$ can be chosen such that $\omega^{ij}_{\ \ \mu} = 0$ and $e_i^\mu = \delta_i^\mu$ at P.* The important consequence of this statement is that it holds not only in GR (i.e. V_4), but also in Poincaré Gauge Theory (i.e. U_4).[130, 131] The EP is not violated in manifolds with torsion, fitting in natural way into a U_4 geometry of space-time. It holds in V_4, as well as in T_4. Note, however, that in more general geometries, characterized by a symmetry of the tangent space higher than the Poincaré group, the usual form of the EP can be violated, and local physics differs from Special Relativity.[129, 132]

3. The Violation of Equivalence Principle in QFT

The EP violation can be achieved in the framework of QFT and GR[133, 134] as well as in the framework of modified gravity[135, 136] and in connection with the generalized uncertainty principle[137]). Consider an electron with mass in thermal equilibrium with a photon heat bath (m_0 is the renormalized mass of the particle at zero temperature). We evaluate the electron gravitational and inertial mass in the low-temperature limit, $T \ll m_0$. We will see that $m_g \neq m_i$ at $T \neq 0$, while $m_g = m_i$ at $T = 0$.

The gravitational and inertial masses are derived by means of the Foldy–Wouthuysen transformation[138] on the Dirac equation (it allows to

---

[i]In this Section, we follow the notation adopted by Blagojevic.[129] A space $(L_4; g)$ with the most general metric compatible linear connection Γ is called Riemann–Cartan space U_4. If the torsion vanishes, a U_4 becomes the Riemannian space V_4 of GR; if, alternatively, the curvature vanishes, a U_4 becomes the Weitzenbock teleparallel space T_4. The condition $R^\alpha_{\beta\gamma\chi} = 0$ transforms a V_4 into a Minkowski space M_4, and $T^\alpha_{\beta\gamma} = 0$ transforms a T_4 into an M_4.

write the Dirac equation in a Schrödinger equation form). To operationally define the inertial mass, an electric field can be applied to charged particles and the consequent acceleration is derived.[133, 134] Taking into account the finite temperature contributions, one obtains (after the renormalization procedure)[133, 134]

$$\left(\not{p} - m_0 - \frac{\alpha}{4\pi^2}\not{I}\right)\psi = e\Gamma_\mu A^\mu \psi. \tag{30}$$

Here, $\not{p} \equiv \gamma^\mu p_\mu$, $\alpha = 1/137$ is the fine-structure constant, γ^μ are the Dirac matrices, A^μ is the electromagnetic four-potential, and finally

$$I_\mu = 2\int d^3k \frac{n_B(k)}{k_0} \frac{k_\mu}{\omega_p k_0 - \mathbf{p}\cdot\mathbf{k}}. \tag{31}$$

Here, $k_\mu = (k_0, \mathbf{k})$, ω_p and $\mathbf{p}$ are related as $\omega_p = \sqrt{m_0^2 + |\mathbf{p}|^2}$, and $n_B(k)$ is the Bose–Einstein distribution $n_B(k) = (e^{\beta k} - 1)^{-1}$, where $\beta = 1/k_B T$ (k_B is the Boltzmann constant). In Γ_μ, the finite temperature corrections are included in the electromagnetic vertex

$$\Gamma_\mu = \gamma_\mu \left(1 - \frac{\alpha}{4\pi^2}\frac{I_0}{E}\right) + \frac{\alpha}{4\pi^2}I_\mu. \tag{32}$$

The Foldy–Wouthuysen transformation allows to cast (30) in the form $i\frac{\partial\psi_s}{\partial t} = H\psi_s$, with

$$H = m_0 + \frac{\alpha\pi T^2}{3m_0} + \frac{|\mathbf{p}|^2}{2\left(m_0 + \frac{\alpha\pi T^2}{3m_0}\right)} + e\phi + \frac{\mathbf{p}\cdot\mathbf{A} + \mathbf{A}\cdot\mathbf{p}}{2\left(m_0 + \frac{\alpha\pi T^2}{3m_0}\right)} + \cdots \tag{33}$$

The inertial mass can be inferred by computing the acceleration

$$\mathbf{a} = -[H, [H, \mathbf{r}]] = \frac{e\mathbf{E}}{m_0 + \frac{\alpha\pi T^2}{3m}}. \tag{34}$$

This equation allows to identify the inertial mass

$$m_i = m_0 + \frac{\alpha\pi T^2}{3m_0}, \tag{35}$$

from which it follows that the inertial mass m_i of an electron at $T \neq 0$ and m_0 at $T = 0$ differs owing to the radiative corrections (see Eq. (35)). It is worth noting that m_i increases with T as expected since inertia increases when particles travel through the background heat bath. Referring to Donoghue,[133, 134] a similar calculation, in the weak field approximation, leads to the the gravitational mass m_g (see Eq. (1)). The Dirac equation in the gravitational interaction reads[133, 134]

$$\left(\not{p} - m_0 - \frac{\alpha}{4\pi^2} I \right) \psi = \frac{1}{2} h_{\mu\nu} \tau^{\mu\nu} \psi, \tag{36}$$

where $h_{\mu\nu} = 2\phi_g \, \text{diag}(1,1,1,1)$ (as usual, ϕ_g is the gravitational potential) and $\tau^{\mu\nu}$ is the renormalized stress–energy tensor. The Foldy–Wouthuysen transformation allows to derive the Schrödinger-like equation $i\dfrac{\partial \psi_s}{\partial t} = H_g \psi_s$, with

$$H_g = m_0 + \frac{\alpha \pi T^2}{3 m_0} + \frac{|\mathbf{p}|^2}{2 \left(m_0 + \frac{\alpha \pi T^2}{3 m_0} \right)} + \left(m_0 - \frac{\alpha \pi T^2}{3 m_0} \right) \phi_g, \tag{37}$$

The calculation of the acceleration (similar to (34)), yields

$$\mathbf{a} = -[H_g, [H_g, \mathbf{r}]] = \frac{m_0 - \alpha \pi T^2 / 3 m_0}{m_0 + \alpha \pi T^2 / 3 m_0}, \tag{38}$$

so that the gravitational mass is identified as

$$m_g = m_0 - \frac{\alpha \pi T^2}{3 m_0}. \tag{39}$$

Therefore, as expected, $m_g = m_i$ at $T = 0$, so that the difference arises from radiative corrections. From (35) and (39), one gets

$$\frac{m_g}{m_i} = 1 - \frac{2\alpha \pi T^2}{3 m_0^2}, \tag{40}$$

in the first-order approximation in T^2. This result yields a violation of the EP in an Eötvös-type experiment. At temperature of the order $T \sim 300K$, the correction is of the order $\sim 10^{-17}$. The result given in (40) follows from the fact that Lorentz invariance, in the framework of the finite temperature vacuum, is broken so that one can define an absolute motion through the vacuum (at rest with the heat bath).

3.1. *The violation of Equivalence Principle via modified geodesic equation*

An interesting approach to derive the previous results has been proposed by Gasperini.[139] Consider a charged test particle with mass m_0 in thermal equilibrium with a photon heat bath. In the low-temperature limit $T \ll m_0$, the dispersion relation is given by $E = \sqrt{m_0^2 + |\mathbf{p}|^2 + \frac{2}{3}\alpha\pi T^2}$.[133] Following Refs. 133, 139, one arrives to the generalized energy–momentum tensor in curved space-times

$$\Xi^{\mu\nu} = T^{\mu\nu} - \frac{2}{3}\alpha\pi\frac{T^2}{E^2}e^{\mu}_{\hat{0}}e^{\nu}_{\hat{0}}T^{\hat{0}\hat{0}}, \tag{41}$$

where $e^{\mu}_{\hat{0}}$ denotes the vierbein field and the hatted indexes are the ones related to the flat tangent space. The Einstein field equations assume the form $G^{\mu\nu} = \Xi^{\mu\nu}$. From the Bianchi identity $\nabla_{\nu}G^{\mu\nu} = 0$, one gets

$$\nabla_{\nu}T^{\mu\nu} = \nabla_{\nu}\left(\frac{2}{3}\alpha\pi\frac{T^2}{E^2}e^{\mu}_{\hat{0}}e^{\nu}_{\hat{0}}T^{\hat{0}\hat{0}}\right), \tag{42}$$

from which (using $\dot{x}^{\mu} \equiv dx^{\mu}/ds$ and $E = m\dot{x}^{\hat{0}} = m\dot{x}^{\rho}e^{\hat{0}}_{\rho}$) one derives the modified geodesic equation[139]

$$\ddot{x}^{\mu} + \Gamma^{\mu}_{\alpha\nu}\dot{x}^{\alpha}\dot{x}^{\nu}$$

$$= \frac{2}{3}\alpha\pi T^2\left[\frac{\dot{x}^{\nu}\partial_{\nu}e^{\mu}_{\hat{0}}}{mE} - \frac{e^{\mu}_{\hat{0}}\left(\ddot{x}^{\nu}e^{\hat{0}}_{\nu} + \dot{x}^{\nu}\dot{x}^{\beta}\partial_{\beta}e^{\hat{0}}_{\nu}\right)}{E^2} + \frac{\Gamma^{\mu}_{\alpha\nu}e^{\alpha}_{\hat{0}}e^{\nu}_{\hat{0}}}{m^2}\right]. \tag{43}$$

This equation generalizes the geodesic equation for non-vanishing temperature. We shall refer Eq. (43) to the Schwarzschild geometry

$$g_{\mu\nu} = \text{diag}\left(e^{\nu}, -e^{\lambda}, -r^2, -r^2\sin^2\theta\right), \tag{44}$$

where

$$e^{\nu} = e^{-\lambda} = 1 - 2\phi = 1 - \frac{2M}{r}.$$

We confine ourselves to the case of radial motion ($\dot{\vartheta} = \dot{\varphi} = 0$). As a result Eq. (43) leads to (in the weak-field approximation)

$$\ddot{r} = -\frac{M}{r^2}\left(1 - \frac{2\alpha\pi T^2}{3m^2}\right).$$
(45)

To first-order approximation in T^2 gives

$$\frac{m_g}{m_i} = 1 - \frac{2\alpha\pi T^2}{3m_0^2},$$

which coincides with the result (40).

3.2. *Application to the Brans–Dicke metric*

As discussed above, the Brans–Dicke theory[15] is the first alternative to GR which tries to incorporate the Mach principle. In this theory, the gravitational constant is assumed variable (more precisely, it corresponds to a scalar field non-minimally coupled to geometry). This approach represents a satisfactory implementation of Mach's principle than GR.[15–17] Considering the above Brans–Dicke action (20), the field equations read

$$2\varphi G_{\mu\nu} = T_{\mu\nu} + T_{\mu\nu}^\varphi - 2\left(g_{\mu\nu}\nabla_\mu\nabla_\nu\right)\varphi,$$
(46)

$$\Box\varphi = \zeta^2 T,$$
(47)

where $\zeta^{-2} = 6 + 4\omega$ and $T = g^{\mu\nu}T_{\mu\nu}$ is the trace of the stress-energy tensor. For a static and isotropic space-time, the line element is

$$ds^2 = e^v dt^2 - e^u\left[dr^2 + r^2\left(d\vartheta^2 + \sin^2\vartheta d\Phi^2\right)\right].$$
(48)

The field equations provide the Brans–Dicke solutions

$$e^v = e^{2\alpha_0}\left(\frac{1 - \frac{B}{r}}{1 + \frac{B}{r}}\right)^{\frac{2}{\lambda}}, \quad e^u = e^{2\beta_0}\left(1 + \frac{B}{r}\right)^4\left(\frac{1 - \frac{B}{r}}{1 + \frac{B}{r}}\right)^{\frac{2(\lambda - C - 1)}{\lambda}},$$
(49)

$$\varphi = \varphi_0\left(\frac{1 - \frac{B}{r}}{1 + \frac{B}{r}}\right)^{-\frac{C}{\lambda}},$$
(50)

Table 5.1. This table includes expected bounds on the parameter ω from different experiments (see[143] and references therein).

Detector	System	Expected bound on ω
aLIGO	$(1.4 + 5)\, M_\odot$	~ 100
Einstein Telescope	$(1.4 + 5)\, M_\odot$	$\sim 10^5$
Einstein Telescope	$(1.4 + 2)\, M_\odot$	$\sim 10^4$
eLISA	$(1.4 + 400)\, M_\odot$	$\sim 10^4$
LISA	$(1.4 + 400)\, M_\odot$	$\sim 10^5$
DECIGO	$(1.4 + 10)\, M_\odot$	$\sim 10^6$
Cassini	Solar System	$\sim 10^4$

with α_0, β_0, B, C, λ, and φ_0 constants. They are connected to the free parameter of the theory ω. From the previous analysis (see (45)), one infers

$$
\ddot{r} = -\frac{v' e^{-u}}{2} \left\{ 1 + \left(e^{-v} - 1 \right) \left(\frac{\lambda B}{r} - C \right) \right.
$$
$$
\left. - \frac{2\alpha \pi T^2}{3m^2} \left[1 + v - \left(\frac{\lambda B}{r} - C \right) v \right] \right\}. \tag{51}
$$

Note that in (51), the correction to the ratio m_g/m_i appears as well as the contribution depending on ω. The latter vanish in the limit $\omega \to \infty$ (i.e. GR is recovered). Using[140] the bound $|(m_g - m_i)/m_i| < 10^{-14}$, the parameters in Ref. 141, and referring to the Earth values, $M_\oplus = 5.97 \cdot 10^{24}\,\mathrm{Kg}$; $R_\oplus = 6.37 \cdot 10^6\,\mathrm{m}$, one gets[142]

$$
\omega > 1.40 \cdot 10^5, \tag{52}
$$

that is similar to a bound recently obtained,[110] which gives $\omega > 3 \cdot 10^5$. In the table, we report the bounds for ω referred to various experiments.[143]

4. The Strong Equivalence Principle in Modified Theories of Gravity

In this section, we shortly discuss the SEP in theories of gravity beyond GR.[144] As mentioned above, modified gravity or ETGs can include, with respect to GR, extra degrees of freedom. These can be scalar, vector, or tensor fields, higher-orders terms in the curvature and torsion invariants, and so on.[1] In these models, such new degrees of freedom couple non-minimally with scalar curvature or scalar torsion. This is the case, for example, of the

Brans–Dicke theory which represents the prototype of scalar–tensor theories. The scalar field φ, in fact, couples minimally to scalar curvature R. The non-minimal coupling generates new (gravitational) interactions among masses. This gives rise to a different modification of the values of the metric perturbations (weak-field approximation) $h_{00} = -2GM_G/r$, which is related to the gravitational mass M_G, and h_{ij}, which is related[j] to the inertial mass M_I,[145,146]

$$M_I = \frac{1}{16\pi G} \int d^4x \sqrt{-g}(h^i_{i,j} - h^i_{j,i})dS^j. \tag{53}$$

In GR, owing to the fact that $h_{ij} = -2GM_G/r$, hence $h_{00} = h_{ij}$, one gets $M_I = M_G$. In Brans–Dicke theory (and more generally, in theories of gravity beyond GR), since $h_{00} \neq h_{ij}$ (weak-field limit of (49)), one infers $M_G = M_I + f(\omega, E_\phi)$, where $f(\omega, E_\phi)$ depends on the parameter ω and the self-energy of the scalar field E_ϕ.[145,146] This fact leads to the violation of SEP[145,146] and the possibility to set up experiments to test it.

5. The Einstein Equivalence Principle and the Black Hole Shadow

Recently,[147] a method has been proposed to test the EEP from electromagnetism by observing the photon ring of black holes. The starting point is a general Lagrangian density that characterizes the EEP violation

$$\mathcal{L}_{em} = -\frac{1}{4}F_{\mu\nu}F^{\mu\nu} - \frac{1}{8}qQ^{\mu\nu\rho\sigma}(x)F_{\mu\nu}F_{\rho\sigma} - eJ^\mu A_\mu, \tag{54}$$

where $F_{\mu\nu} = \nabla_\mu A_\nu - \nabla_\nu A_\mu$ is the standard electromagnetic tensor. Here, again, ∇_μ is the covariant derivative with respect to the Levi-Civta connection, and the last term is the coupling of the electromagnetic field with the matter current J^μ (matter action is not included being irrelevant for the results). The second term describes an unknown background field $Q^{\mu\nu\rho\sigma}$ non-minimally coupled to electromagnetic tensor through the coupling constant q. It represents a possible term beyond the standard physics. The field $Q^{\mu\nu\rho\sigma}$ satisfies the relations $Q^{[\mu\nu]\rho\sigma} = Q^{\mu\nu[\rho\sigma]} = Q^{\mu\nu\rho\sigma}$ and the exchanging symmetry of $\mu\nu$, $\rho\sigma$ as a whole. The quantity $Q^{\mu\nu\rho\sigma}(x)$ could be, in principle, a scalar, vector, tensor, or a sum of these objects. For

---

[j]Here we define the inertial mass as generated in the weak field limit.

example, assuming

$$Q^{\mu\nu\rho\sigma} = 2f(\phi)g^{\rho[\mu}g^{\nu]\sigma},\tag{55}$$

hence $Q^{\mu\nu\rho\sigma}$ is a scalar field ϕ.[148, 149] The variation of the Lagrangian (54) with respect to A_μ gives

$$\nabla_\mu\left[(1 - qf(\phi))\,F^{\mu\nu}\right] = eJ^\nu.\tag{56}$$

which is a modified Maxwell equation. If $\phi(x)$ varies slowly, one gets a result that is equivalent to replacing the electric charge e with a field

$$e' = \frac{e}{1 - qf(\phi)},\tag{57}$$

which means that the fine structure constant has a variable value over the space-time. This result corresponds to the violation of the local position invariance, which is one of the elements of the EEP.[150–152] To experimentally test such a term, one could detect the atomic spectra at different locations (for example, considering an atom on Earth and the same kind of atom on stars in orbits around supermassive black holes[153–155]).

In Ref. 147, the focus is on the scalar part of $Q^{\mu\nu\rho\sigma}(x)$. The geometric optics approximation yields the modified dispersion relation corresponding to the above Lagrangian. For planar circular orbits, photons with different polarizations feel different strength of gravitational potential, inducing a violation of WEP. The observable is expressed as

$$\Delta X = 3\sqrt{3}\Delta\beta\,(u_0),\tag{58}$$

which does depend on the difference $\Delta\beta = \beta^l - \beta^m$, where β is the correction function for the Schwarzschild black hole induced by the violation of EEP. Formally, it is $\beta = \sum_n \beta_n^s u^n$, with $u = 1/r$. Here, $u_0 = 1/3$ represents the radius of the unstable circular orbits of photons in the standard Schwarzschild solution. Equation (58) is the main result;[147] it shows that the difference in the photon ring size of two different linear polarized photons is proportionally connected to the difference of the corresponding EEP violation parameters.

For rotating black holes, results show that a large deviation ratio only occurs when the rotation of black hole is fast and the inclination angle of the rotation axis is nearly edge-on or face-on. As for the Schwarzscshild case, a similar expression of (58) can be derived and (numerically) studied for a small magnitude of the EEP violation and a small rotation speed of black holes.

The general conclusion is that, at the moment, one should be able to distinguish photon rings in the photos of supermassive black holes, which the current observational capabilities do not allow.[156] Recently, it has been shown that the circular photon ring would manifest itself as a periodic visibility function on long interferometric baselines.[157] This fact makes the photon ring distinct in the accretion and the related lensing background. The extension to any shape of the photon ring has been provided,[158, 159] and the corresponding polarimetric signatures on long interferometric baselines has also been studied.[160] Finally,[161] the forecast has been made that a space-based interferometry experiment can reach an accuracy level where the photon ring is insensitive to the astronomical source profile. This is a crucial result since it can be used to precisely test gravity.

It is also worth mentioning[162] that it is possible to show that an observable quantity of the physics around black holes is represented by the two-point correlation function related to the intensity fluctuations on the photon ring.

The next generation Event Horizon Telescope could be able to do the double band observations as well as the corresponding dual-polarizations,[163, 164] allowing to explore possible new physics in the framework of strong gravitational field.

6. Conclusions

GR and metric theories of gravity rely on the validity of the EEP. The latter asserts that the gravitational acceleration is (locally) indistinguishable from acceleration caused by mechanical (apparent) forces. The consequence of EP is that $m_g = m_I$, that is the gravitational mass is equal to inertial mass. The equivalence of inertial and gravitational mass was first pointed out by Galileo and Newton. Einstein, however, was able to recognize this equivalence as a fundamental aspect in which accelerations and forces are also involved, elevating it to a principle. EP led Einstein to formulate a theory in which gravity and acceleration appear on the same physical level, and allowed to postulate that in a free-falling frame, all non-gravitational laws of physics behave as if gravity was absent. EP hence asserts that, in a gravitational field, objects with different (internal) compositions are subject to the same acceleration. This new principle led Einstein to the revolutionary interpretation of gravitation: gravity can be described as a curvature effect of space-time.

As we have seen, the EP encodes the *local Lorentz invariance* (clock rates are independent of the clock velocities), the *local position invariance* (the universality of red-shift), and the *universality of free fall* (all free-falling point particles follow the same trajectories independently of their internal structure and composition). The Lorentz and position invariance are related to the local properties of physics, so that they can be tested by using atomic clocks and measurements of spectroscopy. The universality of free-falling point-like particles, instead, can be tested by tracking trajectories.

The EP can be formulated in two different forms: the weak and the strong form. The weak form of the EP states that the gravitational properties of the interaction of particle physics of the Standard Model (strong and electro-weak interactions) obey the EP. The equality $m_g = m_i$ implies that different (and neutral) test particles undergo the same free fall acceleration, and in a free-falling inertial frame only tidal forces may appear. The strong EP extends the weak one with the inclusion of the gravitational energy. In GR, the strong EP is fulfilled thanks to the gravitational stress-energy tensor, while it can be violated in some extensions of GR. This is the case, for example, of ETGs in which a scalar field is introduced and can be non-minimally coupled with geometry.

The EP has been experimentally verified to remarkable accuracy and nowadays it is still under investigation. Till now experiments have determined upper limits to the differential acceleration $|a_1 - a_2|$ between two freely falling test masses of different compositions. Possible violations of WEP are then quantified by the Eötvös parameter

$$\eta = 2 \left| \frac{a_1 - a_2}{a_1 + a_2} \right|. \tag{59}$$

Tests with increasing accuracy correspond to decreasing upper limits on η. Various kinds of null experiments are possible to test WEP, differing in the magnitude of the potential signal and in the impact of noise sources and systematic effects.

The experimental developments obtained so far allow to get stringent bound on WEP. Several experiments are very active in this moment, as well as future experiments that, thanks to their sensitivity, will allow to set better bounds on η.[151, 165–167] For example, we have

- MICROSCOPE: $\eta < 10^{-15}$
- Lunar Laser Ranging: $\eta < 10^{-14}$
- Moscow Eötvös: $\eta < 10^{-12}$
- Princeton, Fifth-force: $\eta < 10^{-11}$

It is worth mentioning that, by refining the experimental sensitivity, it will be possible to access more stringent constraints also for ETGs and modified gravity. Such technological improvements may be available in the next few years.[168] Such studies may have a relevance not only for curvature-based extensions of GR, but also for affine theories based on torsion invariants[169, 170] and non-metric theories.[171] A possible detection of EP violation offers hence a new possibility to discriminate among concurring theories of gravity.

Acknowledgment

The authors thank the Istituto Nazionale di Fisica Nucleare (INFN), *iniziative specifiche* QGSKY, and MoonLight2 for the support.

A.1. Conformal Transformations in Higher-Order Gravity

In this Appendix, we show the relation between higher-order theories and the scalar-tensor gravity.[172] The first straightforward generalization of GR is

$$\mathcal{L} = \sqrt{-g}\, f(R). \tag{A.1}$$

The variation with respect to $g^{\mu\nu}$ yields the field equations

$$f'(R)R_{\mu\nu} - \frac{1}{2} f(R)g_{\mu\nu} - \nabla_\mu \nabla_\nu f'(R) + g_{\mu\nu}\Box f'(R) = 0, \tag{A.2}$$

with $f' \equiv df(R)/dR$. Equation (A.2) is a fourth-order field equation (in metric formalism). It is convenient to introduce the new set of variables

$$p = f'(R) = f'(g_{\mu\nu}, \partial_\sigma g_{\mu\nu}, \partial_\sigma \partial_\rho g_{\mu\nu}), \tag{A.3}$$

$$\tilde{g}_{\mu\nu} = p g_{\mu\nu}. \tag{A.4}$$

This choice links the *Jordan frame* variables $g_{\mu\nu}$ to the *Einstein frame* variables $(p, \tilde{g}_{\mu\nu})$, where p is some auxiliary scalar field. The term "Einstein frame" comes from the fact that the transformation $g \to (p, \tilde{g})$ allows to recast Eq. (A.2) in a form similar to the Einstein field equations of GR.

In absence of matter, hence $T_{\mu\nu}^{(m)} = 0$, the Einstein equations are

$$\tilde{G}_{\mu\nu} = \frac{1}{p^2}\left[\frac{3}{2}\,p_{,\mu}p_{,\nu} - \frac{3}{4}\,\tilde{g}_{\mu\nu}\tilde{g}^{\alpha\beta}p_{,\alpha}p_{,\beta} + \frac{1}{2}\tilde{g}_{\mu\nu}\left(f(R) - Rp\right)\right]. \qquad (A.5)$$

These equation can be rewritten in a more attractive way by defining $\varphi = \sqrt{\frac{3}{2}}\ln p$, which implies

$$\tilde{G}_{\mu\nu} = \left[\varphi_{,\mu}\varphi_{,\nu} - \frac{1}{2}\tilde{g}_{\mu\nu}\varphi_{,\sigma}\varphi^{,\sigma} - \tilde{g}_{\mu\nu}V(\varphi)\right], \qquad (A.6)$$

where

$$V(\varphi) = \frac{Rf'(R) - f(R)}{2f'^2(R)}\Big|_{R=R(p(\varphi))}. \qquad (A.7)$$

The curvature $R = R(p(\varphi))$ is inferred by inverting the relation $p = f'(R)$ (provided $f''(R) \neq 0$). The field equation (A.6) can be obtained from the Lagrangian (A.1) rewritten in terms of φ and the tilded quantities

$$\mathcal{L} = \sqrt{-\tilde{g}}\left(\frac{1}{2}\tilde{R} - \frac{1}{2}\tilde{g}^{\mu\nu}\varphi_\mu\varphi_\nu - V(\varphi)\right) \qquad (A.8)$$

which have the same form of Einstein gravity minimally coupled to a scalar field in presence of a self-interaction potential. Equation (A.8) clearly suggests why the set of variables $(\tilde{g}_{\mu\nu}, p)$ is called Einstein frame.[173–175]

A comment is in order. As we have seen, in the vacuum, one can pass from the Einstein frame to the Jordan frame and vice versa. However, in the presence of matter fields, a caution is required since particles and photons have to be dealt in different ways. In the case of photons, their world-lines are geodesics both in the Jordan frame and in the Einstein frame. This is not the case for massive particles since their geodesics in the Jordan frame are no longer transformed into geodesics in the Einstein frame, and vice versa, and therefore, the two frames are not equivalent. The consequence is that the physical meaning of conformal transformations is not straightforward, although the mathematical transformations are, in principle, always possible. These considerations extend to any higher-order or non-minimally coupled theory.

References

1. C. M Will, *Theory and Experiment in Gravitational Physics*. Cambridge, University Press, Cambridge (1993).
2. C. M. Will, *Living Rev. Rel.* **17**, 4 (2014).

3. S. Capozziello and M. De Laurentis, *Phys. Rept.* **509**, 167 (2011).
4. S. Capozziello and G. Lambiase, *New Adv. Phys.* **7**, 13 (2013).
5. N. D. Birrell and P. C. W. Davies, *Quantum Fields in Curved Space*. Cambridge University Press, Cambridge, UK (1984).
6. C. M. Will, Was Einstein right?: Testing relativity at the centenary. *Ann. Phys.* **15**, 19 (2005).
7. C. M. Will. *Living Rev. Rel.* **9**, 3 (2006).
8. N. Yunes and X. Siemens, *Living Rev. Rel.* **16**, 9 (2013).
9. K. Akiyama *et al.*, *Astrophys. J. Lett.* **875**, L1 (2019).
10. K. Akiyama *et al.*, *VI. Astrophys. J. Lett.* **875**, L6 (2019).
11. D. Ayzenberg and N. Yunes, *Class. Quant. Grav.* **35**, 235002 (2018).
12. A. H. Guth, *Phys. Rev. D* **23**, 347 (1981).
13. S. Capozziello and V. Faraoni, *Beyond Einstein Gravity*, Volume 170, Springer, Dordrecht (2011).
14. M. Bondi, *Cosmology*, Cambridge University Press, Cambridge (1952).
15. C. Brans and R. H. Dicke, *Phys. Rev.* **124**, 925 (1961).
16. S. Capozziello, R. De Ritis, C. Rubano and P. Scudellaro, *Riv. Nuovo Cim.* **19**, 1 (1996).
17. D. W. Sciama, *Mon. Not. Roy. Astron. Soc.* **113**, 34 (1953).
18. G. A. Vilkovisky, *Class. Quant. Grav.* **9**, 895 (1992).
19. M. Gasperini and G. Veneziano, *Phys. Lett. B* **277**, 256 (1992).
20. A. A. Starobinsky, *Phys. Lett. B* **91**, 99 (1980).
21. J. P. Duruisseau and R. Kerner, *Class. Quant. Grav.* **3**, 817 (1986).
22. P. Teyssandier and Ph. Tourrenc, *J. Math. Phys.* **24**, 2793 (1983).
23. K. Maeda, *Phys. Rev. D* **39**, 3159 (1989).
24. D. Wands, *Class. Quant. Grav.* **11**, 269 (1994).
25. L. Amendola, A. Battaglia Mayer, S. Capozziello, F. Occhionero, S. Gottlober, V. Muller and H. J. Schmidt, *Class. Quant. Grav.* **10**, L43 (1993).
26. A. Battaglia Mayer and H. J. Schmidt, *Class. Quant. Grav.* **10**, 2441 (1993).
27. H. J. Schmidt, *Class. Quant. Grav.* **7**, 1023 (1990).
28. T. Padmanabhan, *Phys. Rept.* **406**, 49 (2005).
29. E. J. Copeland, M. Sami and S. Tsujikawa, *Int. J. Mod. Phys. D* **15**, 1753 (2006).
30. A. Yu. Kamenshchik, Ugo Moschella and V. Pasquier, *Phys. Lett. B* **511**, 265 (2001).
31. T. Padmanabhan, *Phys. Rev. D* **66**, 021301 (2002).
32. B. A. Bassett, M. Kunz, D. Parkinson and C. Ungarelli, *Phys. Rev. D* **68**, 043504 (2003).
33. K. Freese and M. Lewis, *Phys. Lett. B* **540**, 1 (2002).
34. G. R. Dvali, G. Gabadadze and M. Porrati, *Phys. Lett. B* **485**, 208 (2000).
35. S. Capozziello, V. F. Cardone, E. Piedipalumbo, M. Sereno and A. Troisi, *Int. J. Mod. Phys. D* **12**, 381 (2003).
36. O. Bertolami, Ch. G. Boehmer, T. Harko and F. S. N. Lobo, *Phys. Rev. D* **75**, 104016 (2007).
37. S. Capozziello and G. Lambiase, *Phys. Lett. B* **750**, 344 (2015).

38. S. Capozziello, G. Lambiase, M. Sakellariadou and An. Stabile, *Phys. Rev. D* **91**, 044012 (2015).
39. G. Lambiase, A. Stabile and An. Stabile, *Phys. Rev. D* **95**, 084019 (2017).
40. M. Blasone, G. Lambiase, L. Petruzziello and A. Stabile, *Eur. Phys. J. C* **78**, 976 (2018).
41. L. Buoninfante, G. Lambiase, L. Petruzziello and A. Stabile, *Eur. Phys. J. C* **79**, 41 (2019).
42. J. Ehlers and P. Schneider, Gravitational lensing. In Proceedings, 13th International Conference on General Relativity and Gravitation: Cordoba, Argentina, June 28-July 4, 1992, pp. 2140 (1993).
43. L. M. Krauss, *Astrophys. J.* **596**, L1 (2003).
44. C. Itzykson and J. B. Zuber, *Quantum Field Theory*, McGraw-Hill, New York (1980).
45. A. A. Tseytlin, *Phys. Lett. B* **363**, 223 (1995).
46. W. Siegel. Stringy gravity at short distances (2003).
47. T. Biswas, E. Gerwick, T. Koivisto and A. Mazumdar, *Phys. Rev. Lett.* **108**, 031101 (2012).
48. T. Biswas, A. Mazumdar and W. Siegel, *JCAP* **0603**, 009 (2006).
49. J. Edholm, A. S. Koshelev and A. Mazumdar, *Phys. Rev. D* **94**, 104033 (2016).
50. L. Buoninfante, A. S. Koshelev, G. Lambiase and A. Mazumdar, *JCAP* **1809**(09), 034 (2018).
51. L. Buoninfante, A. S. Koshelev, G. Lambiase, J. Marto and A. Mazumdar, *JCAP* **1806**(06), 014 (2018).
52. L. Buoninfante, G. Lambiase and A. Mazumdar, *Nucl. Phys. B* **944**, 114646 (2019).
53. L. Buoninfante, A. S. Cornell, G. Harmsen, A. S. Koshelev, G. Lambiase, J. Marto and A. Mazumdar, *Phys. Rev. D* **98**, 084041 (2018).
54. N. D. Birrell and P. C. W. Davies, *Phys. Rev. D* **22**, 322 (1980).
55. B. L. Nelson and P. Panangaden, *Phys. Rev. D* **25**, 1019 (1982).
56. L. H. Ford and D. J. Toms, *Phys. Rev. D* **25**, 1510 (1982).
57. L. H. Ford, *Phys. Rev. D* **35**, 2339 (1987).
58. B. Allen, *Nucl. Phys. B* **226**, 228 (1983).
59. K. Ishikawa, *Phys. Rev. D* **28**, 2445 (1983).
60. L. F. Abbott, *Nucl. Phys. B* **185**, 233 (1981).
61. T. Futamase, T. Rothman and R. Matzner, *Phys. Rev. D* **39**, 405 (1989).
62. V. Faraoni, *Phys. Rev. D* **53**, 6813 (1996).
63. V. Faraoni, *Cosmology in Scalar Tensor Gravity*, Springer, Vol. 139, (2004).
64. X.-L. Chen, R. J. Scherrer and G. Steigman, *Phys. Rev. D* **63**, 123504 (2001).
65. S. Sonego and V. Faraoni, *Class. Quant. Grav.* **10**, 1185 (1993).
66. A. A. Grib and E. A. Poberii, *Helv. Phys. Acta* **68**, 380 (1995).
67. A. A. Grib and W. A. Rodrigues, *Grav. Cosmol.* **1**, 273 (1995).
68. S. Deser and R. I. Nepomechie, *Ann. Phys.* **154**, 396 (1984).
69. V. Faraoni and E. Gunzig, *Int. J. Theor. Phys.* **38**, 217 (1999).

70. T. P. Sotiriou, V. Faraoni and S. Liberati, *Int. J. Mod. Phys. D* **17**, 399 (2008).
71. I. L. Buchbinder and S. D. Odintsov, *Sov. Phys. J.* **26**, 721 (1983).
72. I. L. Buchbinder and S. D. Odintsov, *Lett. Nuovo Cim.* **42**, 379 (1985).
73. T. Muta and S. D. Odintsov, *Mod. Phys. Lett. A* **6**, 3641 (1991).
74. E. Elizalde and S. D. Odintsov, *Phys. Lett. B* **333**, 331 (1994).
75. I. L. Buchbinder, S. D. Odintsov and I. L. Shapiro, *Effective Action in Quantum Gravity*. Bristol, UK (1992).
76. M. Reuter, *Phys. Rev. D* **49**, 6379 (1994).
77. A. Barroso, J. Casasayas, P. Crawford, P. Moniz and A. Nunes, *Phys. Lett. B* **275**, 264 (1992).
78. R. Fakir and S. Habib, *Mod. Phys. Lett. A* **8**, 2827 (1993).
79. J. Garcia-Bellido and A. D. Linde, *Phys. Rev. D* **52**, 6730 (1995).
80. E. Komatsu and T. Futamase, *Phys. Rev. D* **58**, 023004 (1998). [Erratum: *Phys. Rev. D* **58**, 089902(1998)].
81. B. A. Bassett and S. Liberati, *Phys. Rev. D* **58**, 021302 (1998) [Erratum: *Phys. Rev. D* **60**, 049902 (1999)].
82. T. Futamase and M. Tanaka, *Phys. Rev. D* **60**, 063511 (1999).
83. D. S. Salopek, J. R. Bond and J. M. Bardeen, *Phys. Rev. D* **40** 1753 (1989).
84. R. Fakir and W. G. Unruh, *Phys. Rev. D* **41**, 1783 (1990).
85. R. Fakir, S. Habib and W. Unruh, *Astrophys. J.* **394**, 396 (1992).
86. J. Hwang and H. Noh, *Phys. Rev. Lett.* **81**, 5274 (1998).
87. V. Faraoni, Superquintessence. *Int. J. Mod. Phys. D* **11**, 471 (2002).
88. S. Tsujikawa, K. i Maeda and T. Torii, *Phys. Rev. D* **61**, 103501 (2000).
89. S. Tsujikawa and B. A. Bassett, *Phys. Lett. B* **536**, 9 (2002).
90. J. J. Halliwell and R. Laflamme, *Class. Quant. Grav.* **6**, 1839 (1989).
91. D. H. Coule and K. Maeda, *Class. Quant. Grav.* **7**, 955 (1990).
92. D. H. Coule, *Class. Quant. Grav.* **9**, 2353 (1992).
93. W. A. Hiscock, *Class. Quant. Grav.* **7**, L235 (1990).
94. J. J. van der Bij and E. Radu, *Nucl. Phys. B* **585**, 637 (2000).
95. Ph. Jetzer, *Phys. Rept.* **220**, 163 (1992).
96. J. J. van der Bij and M. Gleiser, *Phys. Lett. B* **194**, 482 (1987).
97. A. R. Liddle and M. S. Madsen, *Int. J. Mod. Phys. D* **1**, 101 (1992).
98. M. B. Voloshin and A. D. Dolgov, *Sov. J. Nucl. Phys.* **35**, 120 (1982) [*Yad. Fiz.* **35**, 213 (1982)].
99. Ch. T. Hill and D. S. Salopek, *Ann. Phys.* **213**, 21 (1992).
100. Y. Hosotani, *Phys. Rev. D* **32**, 1949 (1985).
101. A. P. Lightman and D. L. Lee, *Phys. Rev. D* **8**, 364 (1973).
102. K. S. Thorne, D. L. Lee and A. P. Lightman, *Phys. Rev. D* **7**, 3563 (1973).
103. C. M. Will, *Phys. Rept.* **113**, 345 (1984).
104. A. Coley, *Phys. Rev. Lett.* **49**, 853 (1982).
105. W.-T. Ni, *Phys. Rev. Lett.* **38**, 301 (1977).
106. H. C. Ohanian, *Phys. Rev. D* **10**, 2041 (1974).
107. A. J. Accioly and U. F. Wichoski, *Class. Quant. Grav.* **7**, L139 (1990).
108. U. F. Wichoski. J. Accioly and N. Bertarello, *Braz. J. Phys.* **23**, 392 (1993).
109. M. P. Haugan, *Ann. Phys.* **118**, 156 (1979).

110. C. M. Will, *Living Rev. Rel.* **9**, 3 (2006).
111. F. A. E. Pirani, Bull. LAcademie Polonaise des Science, *Ser. Sci. Math. Astr. Et. Phys.* **13**, 239–242 (1965).
112. F. A. E. Pirani, Building spacetime from light rays and free particles. In *Symposis Mathematica* **12**, 67 (1973).
113. H. Pfister and M. King, *Inertia and gravitation*, Volume **897**, Springer International Publishing, Cham, 2015.
114. R. H. Dicke, The many faces of mach, In H. Y. Chiu and W. F. Hoffmann, editors, *Gravitation and Relativity*, p. 121, Benjamin, New York (1964).
115. P. Jordan, Schwerkraft und Weltall: Grundlagen der theoretischen Kosmologie. F. Vieweg (1955).
116. R. H. Dicke, *Phys. Rev.* **125**, 2163 (1962).
117. M. B. Green, J. H. Schwarz and E. Witten, *Superstring Theory*. Volume 1: Introduction (1988).
118. A. A. Tseytlin, *Int. J. Mod. Phys. A* **4**, 4249 (1989).
119. J. Scherk and J. H. Schwarz, *Nucl. Phys. B* **81**, 118 (1974).
120. N. A. Chernikov and E. A. Tagirov, *Ann. Inst. H. Poincare Phys. Theor. A* **9**, 109 (1968).
121. C. G. Callan, Jr., S. R. Coleman and R. Jackiw, *Ann. Phys.* **59**, 42 (1970).
122. R. Penrose, Zero rest mass fields including gravitation: Asymptotic behavior. *Proc. Roy. Soc. Lond. A* **284**, 159 (1965).
123. R. M. Wald, *General Relativity*. Chicago University Pr., Chicago, USA (1984).
124. J. O. Dickey *et al.*, *Science* **265**, 482 (1994).
125. J. H. Taylor and J. M. Weisberg, *Astrophys. J.* **253**, 908 (1982).
126. A. Zee, *Phys. Rev. Lett.* **42**, 417 (1979).
127. J. L. Cervantes-Cota and H. Dehnen, *Phys. Rev. D* **51**, 395 (1995).
128. J. L. Cervantes-Cota and H. Dehnen, *Nucl. Phys. B* **442**, 391 (1995).
129. M. Blagojevic, *Gravitation and Gauge Symmetries*, Bristol, UK (2002).
130. P. von der Heyde, *Lett. Nuovo Cimento* **14**, 250 (1975).
131. D. Hartley, *Class. Quant. Grav.* **12**, L103 (1995).
132. F. W. Hehl, J. Dermott McCrea, E. W. Mielke and Y. Neeman, *Phys. Rept.* **258**, 1 (1995).
133. J. F. Donoghue, B. R. Holstein and R. W. Robinett, *Phys. Rev. D* **30**, 2561 (1984).
134. J. F. Donoghue, B. R. Holstein and R. W. Robinett, *Gen. Rel. Grav.* **17**, 207 (1985).
135. L. Hui, A. Nicolis and Ch. Stubbs, *Phys. Rev. D* **80**, 104002 (2009).
136. C. Armendariz-Picon and R. Penco, *Phys. Rev. D* **85**, 044052 (2012).
137. F. Scardigli, G. Lambiase and E. C. Vagenas, *Phys. Lett. B* **767**, 242 (2017).
138. L. L. Foldy and S. A. Wouthuysen, *Phys. Rev.* **78**, 29 (1950).
139. M. Gasperini, *Phys. Rev. D* **36**, 617 (1987).
140. S. Baessler, B. R. Heckel, E. G. Adelberger, J. H. Gundlach, U. Schmidt and H. E. Swanson, *Phys. Rev. Lett.* **83**, 3585 (1999).
141. A. Barros and C. Romero, *Phys. Lett. A* **245**, 31 (1998).

142. M. Blasone, S. Capozziello, G. Lambiase and L. Petruzziello, *Eur. Phys. J. Plus* **134**, 169 (2019).
143. K. G. Arun and A. Pai, *Int. J. Mod. Phys. D* **22**, 1341012 (2013).
144. S. Capozziello and V. Faraoni, *Beyond Einstein Gravity: A Survey of Gravitational Theories for Cosmology and Astrophysics.* Springer, Dordrecht (2011).
145. H. C. Ohanian, *Ann. Phys.* **67**, 648 (1971).
146. H. C. Ohanian, *J. Math. Phys.* **14**, 1892 (1973).
147. C. Li, H. Zhao and Y.-F. Cai, *Phys. Rev. D* **104**, 064027 (2021).
148. H. B. Sandvik, J. D. Barrow and J. Magueijo, *Phys. Rev. Lett.* **88**, 031302 (2002).
149. J. D. Bekenstein, *Phys. Rev. D* **25**, 1527 (1982).
150. C. M. Will, *Living Rev. Rel.* **17**, 4 (2014).
151. G. M. Tino, L. Cacciapuoti, S. Capozziello, G. Lambiase and F. Sorrentino, *Prog. Part. Nucl. Phys.* **112**, 103772 (2020).
152. E. Di Casola, S. Liberati and S. Sonego, *Am. J. Phys.* **83**, 39 (2015).
153. R. Angelil and P. Saha, *Astrophys. J. Lett.* **734**, L19 (2011).
154. A. Amorim *et al.*, *Phys. Rev. Lett.* **122**, 101102 (2019).
155. A. Hees *et al.*, *Phys. Rev. Lett.* **124**, 081101 (2020).
156. S. E. Gralla, *Phys. Rev. D* **103**, 024023 (2021).
157. M. D. Johnson *et al.*, *Sci. Adv.* **6**, 1310 (2020).
158. S. E. Gralla, *Phys. Rev. D* **102**, 044017 (2020).
159. S. E. Gralla and A. Lupsasca, *Phys. Rev. D* **102**, 124003 (2020).
160. E. Himwich, M. M. D. Johnson, A. Lupsasca and A. Strominger, *Phys. Rev. D* **101**, 084020 (2020).
161. S. E. Gralla, A. Lupsasca and D. P. Marrone, *Phys. Rev. D* **102**, 124004 (2020).
162. S. Hadar, M. D. Johnson, A. Lupsasca and G. N. Wong, *Phys. Rev. D* **103**, 104038 (2021).
163. L. Blackburn *et al.*, Studying Black Holes on Horizon Scales with VLBI Ground Arrays, arXiv:1909. 01411 [astro-ph. IM].
164. K. Haworth *et al.*, Studying black holes on horizon scales with space-VLBI, arXiv:1909. 01405 [astro-ph. IM].
165. H. Pihan-Le Bars *et al.*, *Phys. Rev. Lett.* **123**, 231102 (2019).
166. P. Touboul *et al.*, *Class. Quant. Grav.* **36**, 225006 (2019).
167. A. M. Nobili and A. Anselmi, *Phys. Rev. D* **98**, 042002 (2018).
168. A. M. Nobili and A. Anselmi, *Phys. Lett. A* **382**, 2205 (2018).
169. Y.-F. Cai, S. Capozziello, M. De Laurentis and E. N. Saridakis, *Rept. Prog. Phys.* **79**, 106901 (2016).
170. P. Salucci *et al.*, *Front. Phys.* **8**, 603190 (2021).
171. J. B. Jimenez, L. Heisenberg and T. S. Koivisto, *Universe* **5**, 173 (2019).
172. R. Schimming and H.-J. Schmidt, *Gesch. Naturw. Tech. Med.* **27**, 41 (1990).
173. A. Strominger, *Phys. Rev. D* **30**, 2257 (1984).
174. V. Faraoni, *Phys. Rev. D* **74**, 104017 (2006).
175. V. Faraoni and S. Nadeau, *Phys. Rev. D* **75**, 023501 (2007).

www.ingramcontent.com/pod-product-compliance
Lightning Source LLC
Chambersburg PA
CBHW050629190326
41458CB00008B/2192